DONATION

285 4903

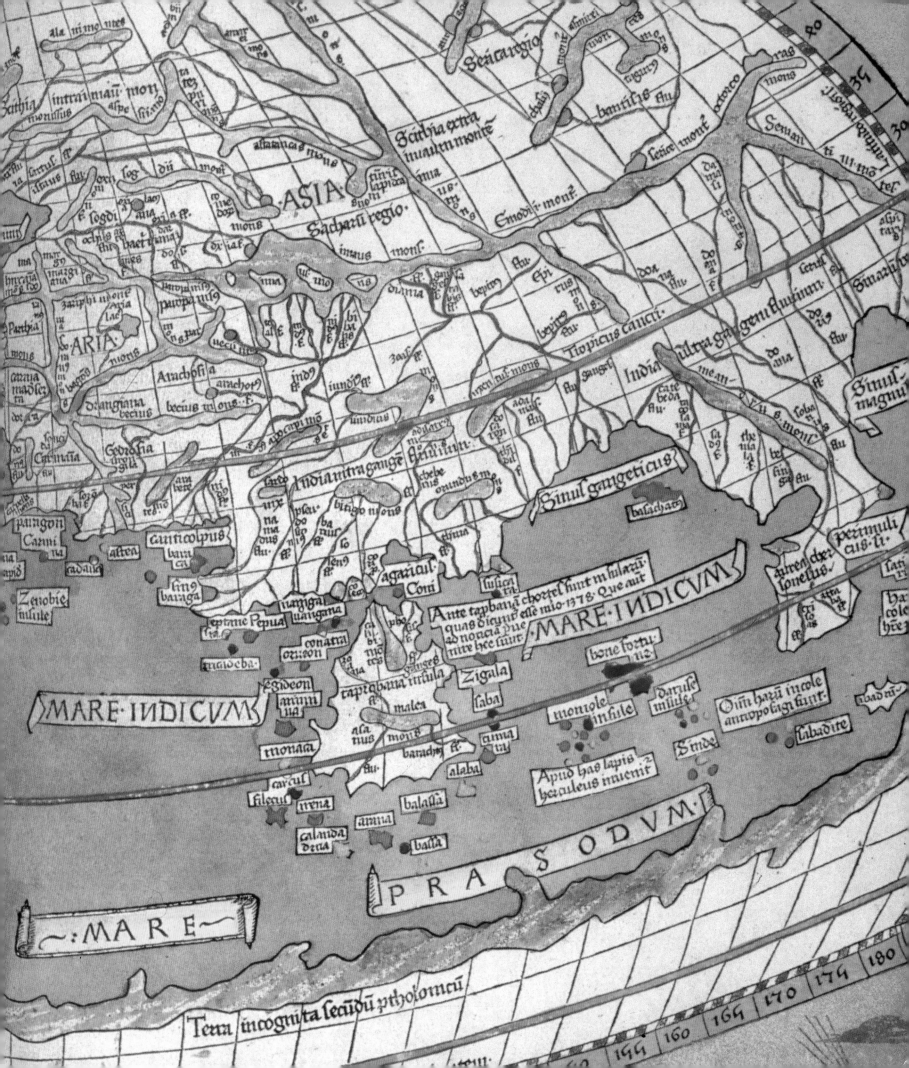

The Image of the World

'Paradise is now shut and locked, barred by angels; so now we must go forward, *around* the world, and see if somehow, somewhere there is a back way in.'

— HEINRICH VON KLEIST 1811

The Image of the World

20 CENTURIES OF WORLD MAPS

Peter Whitfield

THE BRITISH LIBRARY

FRONT ENDPAPERS
Ptolemaic World Map, 2nd century AD, republished 1486, British Library Maps, 1.d.2

REAR ENDPAPERS
Dynamic Planet World Map, USGS 1989, British Library Maps X 1179

TITLE PAGE
World Map of Antonio Sanches, 1623, British Library Add MS 22874

© 2010 Peter Whitfield

First published 1994
This edition published 2010 by
The British Library
96 Euston Road
London NW1 2DB

British Library Cataloguing in Publication Data
A CIP record for this title is available from The British Library

ISBN 978-0-7123-5089-1

Designed by John Mitchell
Typeset by Bexhill Phototypesetters, Bexhill-on-Sea, East Sussex
Additional layout & typesetting by Andrew Barron@thextension
Printed in Italy by Grafiche Milani

CONTENTS

PREFACE

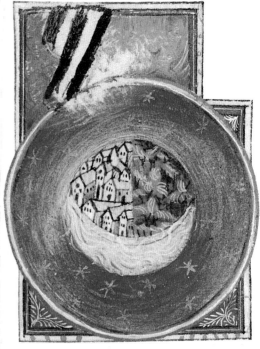

I HAVE WRITTEN THIS BOOK IN an attempt to bring out something of the wider cultural significance of world maps chosen from different historical periods. Writers on cartographic history have often been reluctant to interpret maps, or to relate them to the history of ideas, to art, and to the societies in which they were created. They have consequently understated a most important theme in the understanding of maps. I have argued that the maps of the past have often contained deeply subjective elements, and that a subjective approach to them is therefore a valid way of interpreting what is happening in these maps, and what lies behind them. The act of representing reality in maps may not be too different from the act of representing it in art or literature: the impulse to crystallize, comprehend, and therefore control aspects of reality. In developing this argument I have borrowed a number of terms from literary or artistic usage and applied them to the study of maps, such as tradition and creativity, symbolism and rhetoric. The word *image* occurs repeatedly in the book, a word that has perhaps become imprecise and devalued. It is used to denote the central fact of mapmaking, namely communication, whether communication of data, of ideas or of fantasies. By the word *image* I understand the wholeness of the map — the motives behind its creation, the materials it uses, and the impact it was intended to make on its contemporary audience.

A medieval image of the earth, composed of houses, forest and sea, against a star-filled sky. Early 15th century manuscript of *L'Image du Monde* by Gautier de Metz.

An interpretative book like this could have been written only on the basis of the accurate scholarship of the past, from the work of the nineteenth-century pioneers Jomard and Santarem, to *The History of Cartography* now in progress and works such as R.W.Shirley's invaluable carto-bibliography *The Mapping of the World.* I am most grateful to Tony Campbell and Peter Barber of The British Library and to Professor David Woodward of the University of Wisconsin-Madison who have read and commented on this text and made many valuable suggestions. Peter Barber has researched and described in detail the Evesham Mappa Mundi, and generously lent me his article on the map, forthcoming in *Imago Mundi,* 1995. I have received much research assistance over a number of years from Francis Herbert of the Royal Geographical Society, and more recently from Nick Millea and his colleagues at the Bodleian Library, and from Jeff Armitage and the British Library staff. David Way has admirably fulfilled the role of the Gentle Reader, while Kathleen Houghton secured the illustrations. Robin L. Marsh generously lent a number of maps from his private collection.

A few of the older maps reproduced in this book have survived in unique copies, uncoloured and in rather poor condition. These have been photographically enhanced and coloured in order to re-create something of their original impact.

INTRODUCTION
Outer Worlds and Inner Worlds

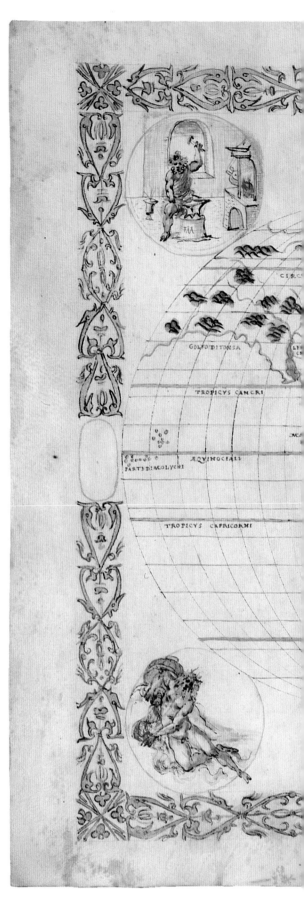

'NOTHING IS DEFINABLE UNLESS IT HAS NO HISTORY.' Thus Nietzsche crushed the ambitions of the cultural historian who claims the ability to rationalize and interpret the past. The world map has a recoverable history of more than two thousand years, and there is no meaningful definition that would cover all the images in this book. The reasons why the world map took the forms that it did, the psychological processes that created it, are ultimately hidden from us. Maps are cultural artefacts, comparable in history perhaps to arms and armour, or musical instruments, or ships. Almost all cultures have developed these things, but with enormously varying degrees of sophistication. Their origin is instinctive, in that they are products of both the intellect and the imagination confronting problems in reality. They have faced severe practical constraints in their construction and their use, but they have evolved because they were of fundamental importance. They have acquired an aesthetic dimension, and their forms have been influenced by art, imagination, symbolism as well as by empirical knowledge. The historian can take account of these interacting forces and try to analyse the way in which they have shaped the evolving world map. But the history of mapping, like the history of most things, is not a science: it can describe but not ultimately explain.

The technical means by which the earth has been explored and measured is one aspect of the history of maps. In one sense this book is clearly an account of geographical progress, of the dialogue between mapmakers and reality. The most striking fact of that dialogue is that the world map represents a reality which, although present to our senses, is perpetually out of reach. The scientific means by which that problem has been solved deserves a technical history in its own right. But the world map has always been shaped not by science alone but by religion, politics, art and obsession. Themes such as divine power, the natural elements, secular ambitions, recur constantly and express more than pure geography. These influences have been at times conscious, at times unconscious. Throughout the greater part of history the sources of knowledge lay in inherited authority and beliefs, not in reason or experience, and these sources have left their imprint unmistakably on the world map. Moreover the forms in which even scientific knowledge is expressed are constantly evolving, mirroring the societies from which they spring.

These maps are rooted in the history which they help us to create; therefore they must often be interpreted in language which their contemporaries would not have recognized. This approach carries a danger that haunts all historians of ideas: the temptation to demythologize or decode the past. If everything is culturally relative, nothing is what it seems, and everything must be interpreted by the use of subtle intellectual keys. Such an approach becomes self-defeating if it distances us from the mind of the past, and demonstrates merely the mind of the present. With that danger held clearly in mind, this book attempts to interpret maps as historical documents whose contemporary context is essential to their understanding. In particular the principle of subjectivity in maps, both personal and cultural subjectivity, is emphasized as an analytical key. There is a natural assumption that maps offer objective depictions of the world. The message of this book is that they do not, and that the innumerable ways in which they do not, serve to place maps as central and significant products of their parent cultures.

Subjectivity carries the implication of freedom, imagination, almost perhaps of play — elements whose role in mapmaking have not been sufficiently recognized. The richest and

A manuscript world map by Francesco Ghisolfi, Florence, c.1565.

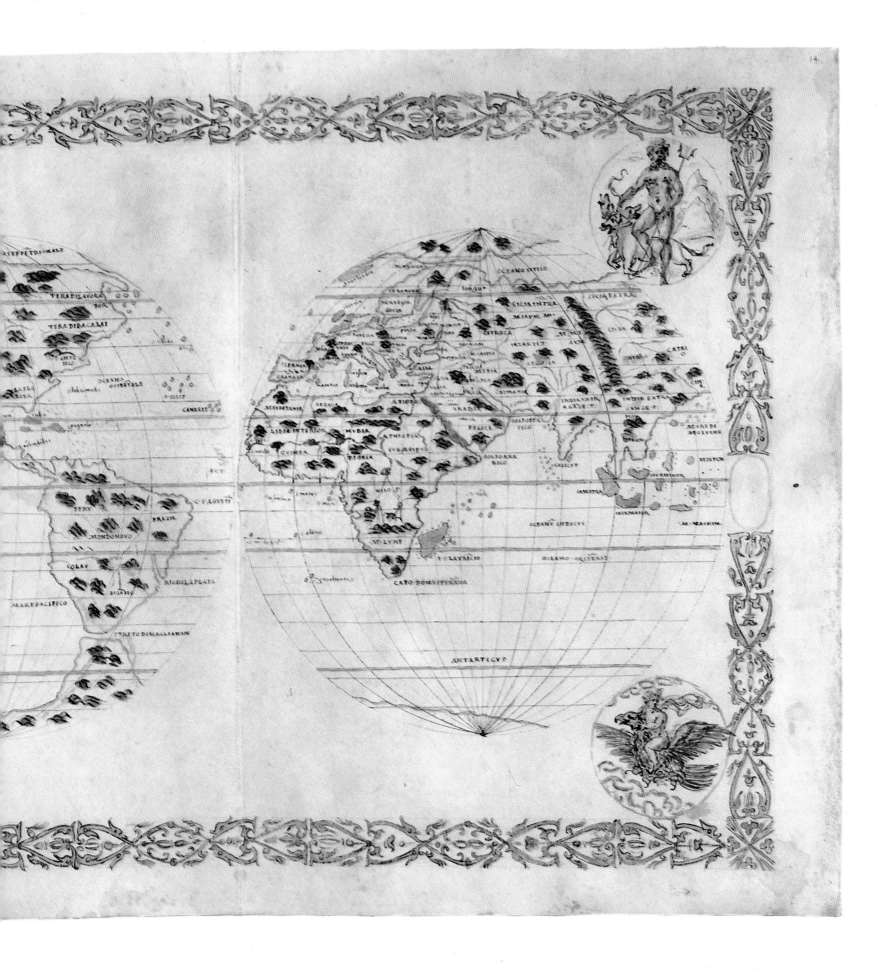

most revealing world maps are those which are least self-conscious. The impulse to depict the world on paper has always been associated with the desire to make some statement about the world. These statements have been intuitive, and frequently expressed in symbolic form, but in essence they are statements about man's belonging in the world, and about his ability to understand or master his environment. The world maps in this book record, inadvertently, the systems of belief and knowledge which enabled him to achieve this. If one had to choose the single most telling motif, constantly recurring in countless world maps over the centuries, it is that of power, the controlling powers that shape the world's features and history. That power may be religious — Christian or pagan. It may be secular — conquest, trade or empire. It may be conceptual — the world map as a navigational instrument or as a thematic document. Or it may be scientific — cosmological or seismic. It is when these themes are unselfconsciously expressed that the world map receives most clearly the intellectual imprint of its time.

It is the sense of not belonging, not recognizing in the received world map the lineaments of one's own age, that has led to perpetual innovation in world mapping, as each age has redefined its sources of knowledge and authority. That knowledge has become progressively more impersonal, setting up a tension with the intuitive, poetic impulse within mapmaking. This is the meaning of Kleist's ironic insight that, expelled from paradise by knowledge and self-consciousness, we must travel around the world, and try to find a back door into Eden. We attempt this, metaphorically, by exploration and by the making of new maps. The great problem with the modern world map, especially since Kleist's day, is that the diversity of knowledge has created a multiplicity of world maps, from which we must choose the image of reality with which we feel at home. We can no longer simply relocate paradise from one part of the world to another, as the medieval mapmakers did.

It is self-evident that the only true and accurate map of the earth is a three-dimensional globe, yet mapmakers have persistently striven to re-create the world on paper. Why? This question goes to the heart of the history of the world map: the desire to see an image of the entire world focused before us, clear, self-contained, comprehensible and masterable. Attempts to resolve this paradox have been distilled over the years in a cartographic language, which like all languages involves symbolism. Sometimes that symbolism has been calculated and lucid, sometimes it has been intuitive and unspoken. The task of the map historian is to describe and interpret the development of that language, and relate it to other forms of creative expression.

The elusive character of maps, their twin roots in reality and imagination, their distorting and yet revelatory quality, were all recognized by one of the great medieval mapmakers, Fra Mauro, and his words are no less valid to the writer of histories than to the maker of maps:

' If anyone considers incredible the unheard-of things I have set down here,
let him do homage to the secrets of nature, rather than consult his intellect.
For nature conceives of innumerable things, of which those known to us are
fewer than those not known, and this is so because nature exceeds understanding.'

From the world map by Pierre Desceliers, 1550, a manuscript map made for royal presentation. The geography of South Asia and the Spice Islands is detailed and accurate, while the decorative figures are pure legend.

THE IMAGE OF THE WORLD

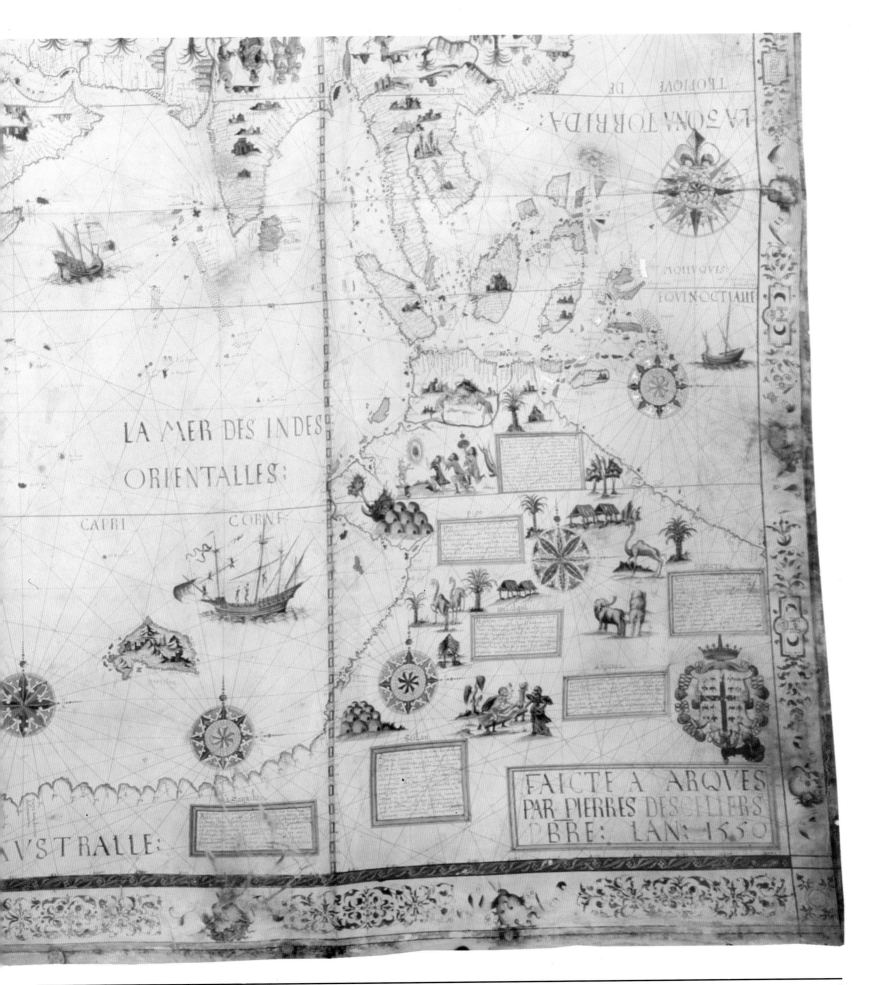

TROPIQVE DE

LA ZONA TORRIDA:

MOUUQUES

EOUINOCTIALE

LA MER DES INDES

ORIENTALLES:

CAPRI   CORNE

FAICTE A ARQVES
PAR PIERRES DESCELLERS
PBRE: L'AN: 1550

AVSTRALLE:

CHAPTER ONE
Classical Foundations

T HE OLDEST SURVIVING IMAGES OF the world are from the civilisations of the ancient near east. These figures of the world typically portrayed such elements as the gods or dominant forces; man with symbols from his environment such as artefacts, weapons or animals; and then the basic natural elements of earth, sky and water. Should these figures be regarded as maps? Perhaps, since their makers took picture-elements from the world around them, and arranged them into coherent diagrams. But they naturally lack any kind of spatial precision, for their purpose was poetic or religious rather than geographical. They served to define man within the other elements of his world, or sometimes one group of men against their neighbours or enemies.

For the purposes of this study the world map crystallizes in Greek thought from the sixth century B.C. onwards. The difficulties of reconstructing the Greek sciences — in this case geography and astronomy — have often been commented on. Many important writers are known only through second- or even third-hand reports, so that we do not have a coherent presentation of their ideas. We have outlines, or suggestions, or guiding principles, and the precise meaning of crucial words and phrases may be elusive. In the case of geography, the overwhelming fact is that no world map in any form has survived from the entire classical period. From major and minor thinkers — Pythagoras, Herodotus, Aristotle, Hipparchus — we have only descriptive texts which must be used to reconstruct fundamental geographical concepts. The modern maps which have appeared in many books with such titles as *The World as Known to Herodotus* or *The World According to Strabo* have their uses, but should always been seen as secondary interpretations or reconstructions.

Ptolemaic World Map, printed in Rome, 1478. Note the differences in geography and projection compared with the Ulm Ptolemy of 1482, pages 8–9.

Left: Picture map from c.3000 B.C. Redrawn from engraved silver vase from the Caucasus region.

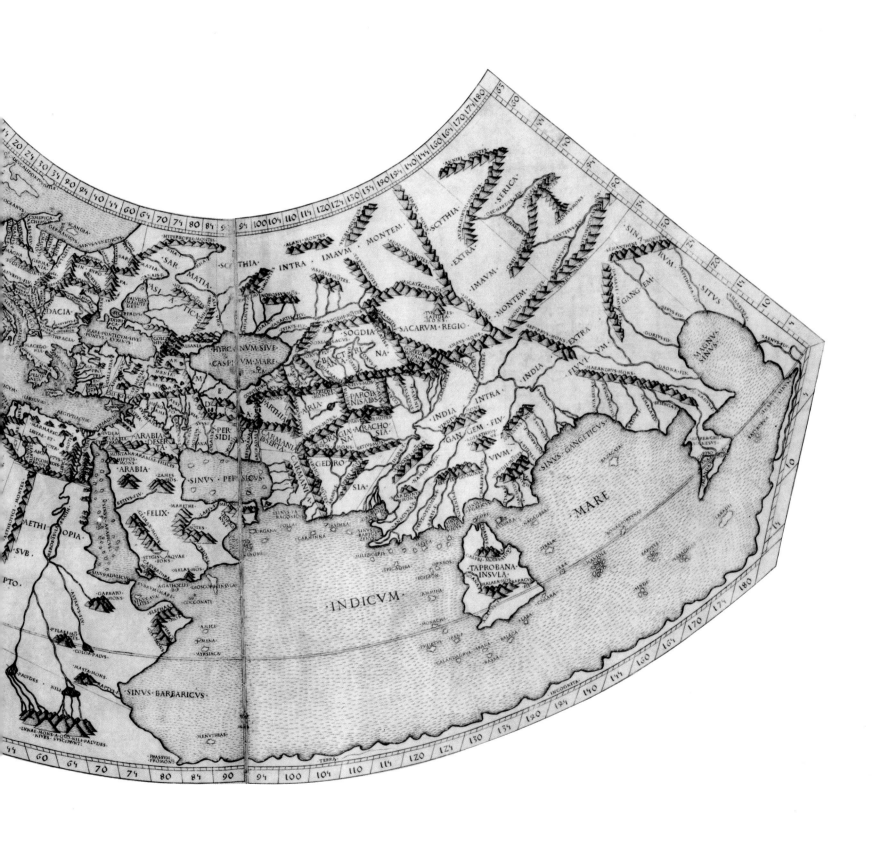

Greek writers from the sixth century B.C. onwards speculated extensively on the shape of the earth and the extent of its inhabited section. They appear to have made an important distinction between the earth as a whole (*gea*), and the known, inhabited world of man (*oikoumene*). Among others, Hecataeus of Miletus (fl. 500 B.C.) described the world as a circular land-mass quite surrounded by water, while his contemporary Anaximander is reported to have constructed a globe. The reports of these two early theorists in the work of Strabo, written five centuries later, illustrate the problem of transferring text into image, compounded when the text is a second-hand report. Did these early thinkers consider the earth to be circular, or the inhabited world? Circular as a disk is circular, or as a sphere? Literary references to circular world maps (e.g. in Herodotus) may signify a two-dimensional representation of a spherical globe. The vexed question of flat earth versus spherical globe has no clear answers until Pythagoras (fl. 500 B.C.) who taught, as a purely theoretical principle, that the world must possess the most perfect form known to nature — the sphere. By contrast Aristotle (fl. 340–320 B.C.) deduced from observation that the world was round, and gave several clear proofs of the fact.

From the third century B.C. onwards, two forces enlarged and clarified the Greek conception of the world. First was progress in the geometry of the sphere, making possible more precise mensuration, position-finding, and ultimately co-ordinate systems. In this respect geometric earth-science followed astronomical theory. The concept of the sky as a sphere, or series of spheres, naturally re-inforced the belief that the centre of such a system, the earth, must in turn be a sphere, and the mapping of the heavens greatly influenced classical methods of earth mapping. Eratosthenes (fl. 240–200 B.C) calculated the circumference of the earth with great accuracy by using spherical geometry, and he devized a system of parallels and meridians for locating places on a world map. Archimedes (fl. 250–220 B.C.) is reported to have made globes and to have constructed models of a planetary system with the earth at its centre — a heliocentric theory had been been advanced by Aristarchus (fl. 250 B.C), but it was universally discounted. Hipparchus (fl. 160–130 B.C) produced the most detailed Greek star catalogue and introduced from Babylonian mathematics the 360 degree division of the globe.

The second source of change in the Greek conception of the world was the extension of the known world through travel, exploration and war. The historian and traveller Herodotus (fl. 450–430 B.C) was critical of geographers who merely theorized about the shape of the world, advocating instead travel and empirical research. He recorded many geographical traditions from beyond Greece, most famously that Africa was navigable, and that a Phoenician expedition had achieved this feat around the year 600 B.C. The voyage of Pytheas c.300 B.C. to the coasts of France, Britain and Germany revolutionized Greek knowledge of northern Europe, indeed many considered his accounts a fabrication. An important reason for this scepticism was the Greek theory of the *climata*: that the earth was naturally divided into latitudinal zones, and that the torrid and frigid zones were quite uninhabitable. Pytheas' accounts of habitation, even civilization, around 60 degrees north, offended against this orthodoxy. The expeditions of Alexander the Great stretched the Greek concept of the world far into the east. The deserts and mountains of south-central Asia were clearly not the limits of the human world, and the circular image of the island world began to look simplistic. The impact of Alexander's journey, together with the concept of the *climata*, produced a map of the world that was trapezoid, approximately twice as broad, east to west, as it was deep, from north to south. Crates (fl. 150 B.C) exhibited in Rome a large model globe, having four land-masses, of which the Eurasian known world was one. The others were evenly distributed north and south of the equator and east and west of an Atlantic meridian, this pattern of continents forming a purely theoretical symmetry. This globe was widely reported and admired. It may be that these geographical ideas were known only to an intellectual elite, but it is equally true that within that elite there was a clear tradition, a body of knowledge and of theory in astronomy, geoscience and maps, that was growing and sharpening throughout the Hellenistic and early Roman period. The pre-Ptolemaic era of Greek geography is summed up by Strabo (fl. 30 B.C. – 10 A.D.), whose writings are also the only source for our knowledge

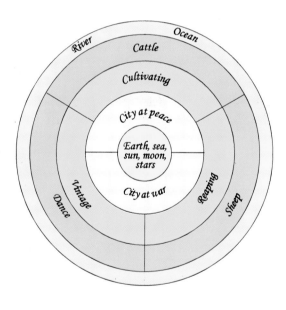

A visualization of the shield of Achilles, described in Homer's *Iliad*. It is thought to represent a type of cosmographical chart, bringing together elements of the natural and human world, although not in a spatial relationship.

Egyptian cosmographical map, c.350 B.C. Stone sar-cophagus lid with the sky-goddess Nut covering the earth. The god of the air, Shu, can be seen below the stars, above the circular earth.

of many earlier Greek geographers. Strabo had no doubt that it was possible to sail around the world eastwards or westwards, for he conceived the inhabited world as an island, located entirely north of the equator, extending to 54 degrees north (to Ireland), and about 125 degrees east to west, from India to Portugal. The possibility that other continents existed was quite accepted, as Crates had shown on his globe, but no time was spent in speculating where they were or what kind of beings inhabited them.

By the first century A.D. Greek-Roman geography formed one intellectual tradition. All the major scientists were Greek, but writing within Roman institutions; owing much to Roman civil and military culture, but representing the development of Greek thought over six centuries. The Greek genius was peculiarly analytical and theoretical, and to this tradition the Romans contributed little if anything. The typical Roman scientist-philosopher was Pliny, the hunter-gatherer of flora, fauna, facts, artefacts, lore and legend, but utterly lacking the analytical impulse. The theoretical spirit reached its culmination in the work of Ptolemy of Alexandria (c.A.D.90–168), who consciously summed up the methods and materials of his predecessors. Ptolemy stood at the end of a line of precocious achievement in classical science. The process by which his work fell into neglect had two main aspects. First, the Roman distaste for theoretical geography; second, the radical discontinuity in learning and literature from the fourth century onwards. Into this vacuum came new geographical ideas supposedly based on the Bible. The latter mark the beginning of the transition to the medieval world map, having quite different sources and purposes from classical maps. This should be seen in the context of the key process in European cultural history from the sixth to the eighth century, that of Christianization; beneath the apparent political chaos of Rome's disintegrating power, the new intellectual empire of medieval Christendom was being formed.

The precise achievement of the classical period is elusive because of the lack of surviving maps. What emerges clearly is the incisive theoretical approach to unravelling the shape of the earth: the world and indeed the universe was felt to be a coherent whole, capable of rational modelling. What was the source of this unique insight? Why did the Greeks create this intellectual framework, and not the Egyptians, the Persians, the Indians? The answer may lie in the simplistic nature of Greek religion. The pantheon of gods, demigods and heroes was little more than a body of folklore, a religion without theology, offering no keys to the mysteries of existence. The questioning intellect was free, indeed was compelled, to seek its own form of truth. Some support for this view appears from the fact that the most theologically mature of all ancient societies — the Jews — lacked any tradition of world mapping or geographical theory. Whatever the true explanation may be, the thinkers of the classical era have bequeathed us, in text rather than in image, the earliest recoverable phase in the history of world mapping.

Ptolemy's World Map, c. 150 A.D. Republished 1482

THE PTOLEMAIC WORLD MAP IS one of the great enigmas in the history of mapping. A unique process of intellectual re-incarnation has given Ptolemy a dual life and made him a pivotal figure in two widely separated periods of history — Hellenistic Greece and Renaissance Italy — so that one must continually remind oneself which Ptolemy one is speaking of. The Greek Ptolemy clearly represented the culmination of six centuries of geographical observation and theory. But the Greek Ptolemy left no surviving maps; his works lay in oblivion in the west for almost a thousand years, and we know him only as a Renaissance scholar, contemporary of Alberti, Brunelleschi and Donatello, offering a rational re-organization of geographical space which no living mapmaker could equal.

The Greek Ptolemy set out consciously to focus, criticize and synthesize the geographical thought of classical Greece. He accomplished this task by addressing three outstanding problems: first, the size and location of the inhabited world (the *oikoumene*); second, the location of specific places upon a world map; third, the mathematical construction of a world map. The achievement of the Greek Ptolemy lies in his recognition and handling of these crucial theoretical issues. On the first question, his originality lies in his cautious restraint: his world is no longer the conceptual island of Crates or Strabo, since all its boundaries are shown to end in unexplored land or sea. The map ends where knowledge ends, and no artificial framework is added. In the context of classical geography this was an innovative step, for the *oikoumene* no longer had fixed geographical boundaries: the possibility of extending the map, especially to the east and south, was obvious. Nor did Ptolemy apparently accept the implications of the *climata* theory, limiting human life effectively to Eurasia and North Africa. There might indeed be inhabited regions outside the known world, he wrote, 'But what these inhabited regions are we have no reliable grounds for saying, for up to now they are unexplored by men from our part of the world'. The most celebrated enigma of the Ptolemaic map is the land linking South-East Asia with Africa. We have no knowledge of Ptolemy's source for this idea; the land is completely featureless, and it may have been added as a purely theoretical balance to the lands north of the equator. Ptolemy's *oikoumene* occupies 80 latitude degrees by 180 longitude degrees, about one quarter of the earth's surface.

On the second question, Ptolemy published effectively the first gazetteer of the world, naming approximately 8,000 places and giving their co-ordinates, which Ptolemy had collated from earlier writers. Longitude is given east from a prime meridian through the Fortunate Isles, latitude from the equator. This gazetteer was the raw material for a series of twenty-seven (more in some manuscripts) detailed regional maps of the known world. Ptolemy was a polymath, the author of works on astronomy, physics, mathematics and optics as well as geography. His treatise on astronomy, the *Almagest* (from its Arabic title *The Master*), is a work of theoretical sophistication and enormous observational detail. In it Ptolemy advances the concept of the heavenly bodies as moving in a series of concentric circles with the earth at the centre. The stars and planets were conceived to be fixed within spheres of invisible crystal. Ptolemy offered the most detailed star catalogue in the ancient world, identifying 1022 individual stars, and locating them by means of a co-ordinate system. It followed logically that places upon the spherical earth should also be

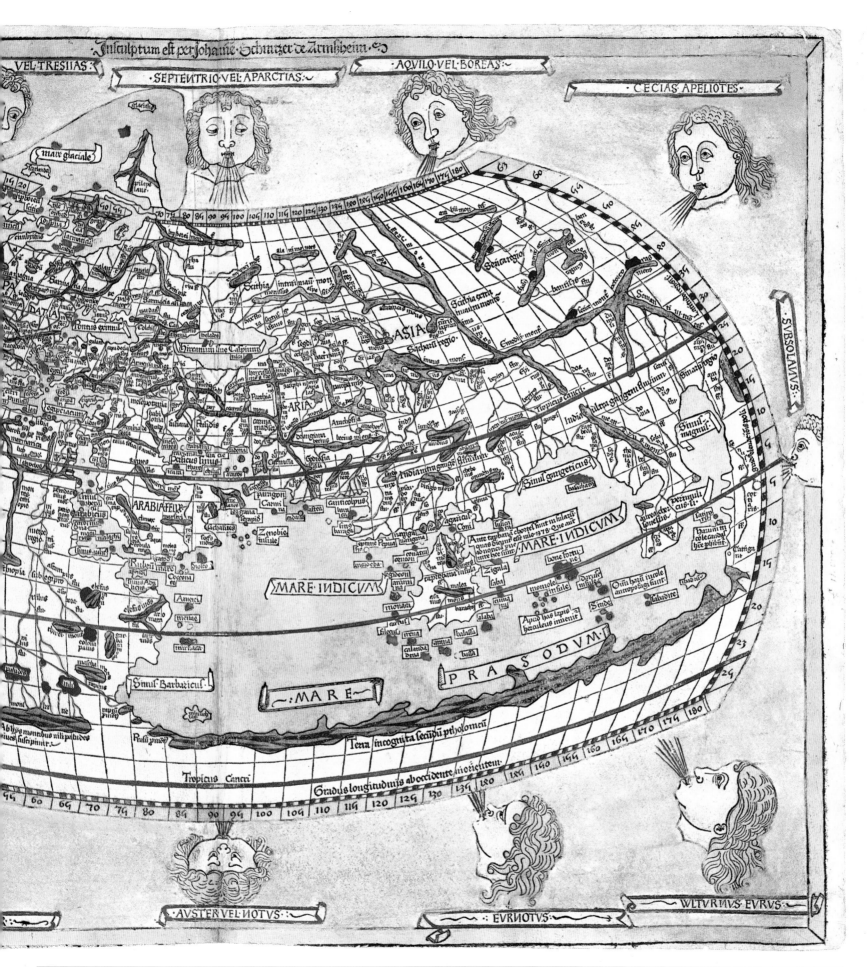

located within an objective framework, that is a geographical co-ordinate system. The rationalizing power of this concept cannot be overemphasized: it is one of the cornerstones of geographical practice, and it was Ptolemy's discovery.

On the third question, Ptolemy elevated cartography onto a new plane by recognizing the problems of transferring a map of the earth's spherical surface onto paper, and by providing three valid solutions to the problem. In his first projection (page 5) the parallels of latitude are arcs of circles, while the meridians are straight lines, deflected at the equator. This has the merit of simplicity in its drafting, but it fails to convey to the eye the sphericity of the earth. The second projection (page 9) curves both the parallels and the meridians, thus achieving the desired optical effect. The third projection, the most complex and least used, was formed around two central axes, both straight lines, with all other meridians and parallels concave to them. This is a sophisticated idea, developed much later for the oval world maps of the Renaissance. Ptolemy's designing of these three complex structures shows a degree of virtuosity that amply justifies his pre-eminence in classical geography.

Having said all this, it would now be natural to examine and interpret these classical world maps. But first we encounter the Ptolemaic paradox: no manuscript of his work is earlier than the thirteenth century, and no contemporary Ptolemaic world map survives. All that we have are reconstructions made from his texts more than a thousand years after his death. As one modern scholar has aptly expressed it 'Ptolemy transmitted his cartographic knowledge in digital rather than in graphic form'. This is particularly important because Ptolemy stood literally at the end of a line of development. Within a century of his death, his ideas, like all Greek learning, were forgotten in the west, while in the eastern empire his texts were preserved but his concepts remained undeveloped. So that although Greek Ptolemy's general achievement is clear, we can only speak of the texts, the maps and the influence of Renaissance Ptolemy. We have no knowledge of how the texts of his works that were published in fifteenth century Italy and Germany may have differed from the originals composed in second-century Alexandria.

After c.300 A.D. there are no references to Ptolemy in Greek or Roman geographical writings. His work was apparently familiar to medieval Arab geographers from the eighth century to Al-Idrisi, but the next documentary evidence we have is that manuscripts of his work were extant in Constantinople in the fourteenth century, having been preserved in both Arab and Byzantine hands during the long period of western neglect. The first Latin translation appeared in Florence in 1406, returning it to the mainstream of European intellectual life, a process that was reinforced by the advent of printing. Printed editions of Ptolemy's *Geography* appeared from 1477 onwards in Rome, Bologna, Florence, Ulm and Strasbourg, and it became the acknowledged fountainhead of geographical science. Some editions slightly modified the world map and added new regional maps to reflect contemporary knowledge, clearly seen for example in the treatment of Scandinavia, while others followed more narrowly their received original.

What then are the key features of Renaissance Ptolemy? What was the flash of recognition that made the fifteenth century scholars receive this map as the world map for their time, against all rival contemporary maps? The crucial concept is that of ordered space. Even the latest and most sophisticated of the circular *mappae mundi*, the Fra Mauro of 1459, appears to have an element of chance, guesswork, almost disorder in its structure. The circular framework was known to be illogical, the sources for its place-names were literary and anecdotal, even legendary, and their location was often arbitrary. Other maps such as the Genoese map of 1457 drew from a store of graphic images which by the late fifteenth century were largely rhetorical. By contrast Ptolemy appeared to have cast a transparent net over the earth's surface, every strand of which was precisely measured and placed. He had defined his subject — one quarter of the earth's surface — and within a geometric framework he had calculated each element of his composition. Moreover the map was only a map, not a visual encyclopedia: descriptive texts do not weigh down the page, and the graphic medieval cavalcade of myth and legend is rested. A dispassionate sense of geographic reality reigns. This sense of ordered space was precisely the ideal towards which the artists of fifteenth century Italy were striving, and this identity of interest explains Ptolemy's appeal. The wind-faces that became such a hallmark of the Ptolemaic map have no warrant in the text; the twelve winds were identified by Aristotle, but the Renaissance-style heads first appear in fifteenth-century maps.

It is easy to see why the fifteenth-century mind so willingly received this revived classicism. In the methodological sense, Ptolemaic map-making deserves its pre-eminence. But it is also true that in the material sphere, the sphere of geographical information, the Ptolemaic revival was a retrograde force, for its world view was essentially that of the Roman Imperial period. It is one of the great ironies of map history that after centuries of neglect, the Ptolemaic map should have been rediscovered at the very moment when contemporary events would reveal its limitations. The scientific aspects of Ptolemy's map — the concept of projection and the use of the graticule — were enduring legacies, but the world map itself was an anachronism, and the cult of Ptolemy was more a literary, intellectual phenomenon than a scientific one.

CLAVDII PTHOLOMEI ALLEXANDRINI COSMOGRAPHI

Ptolemy as the geographer of the Old World: detail from Waldseemüller's world map of 1507 (pages 48–49).

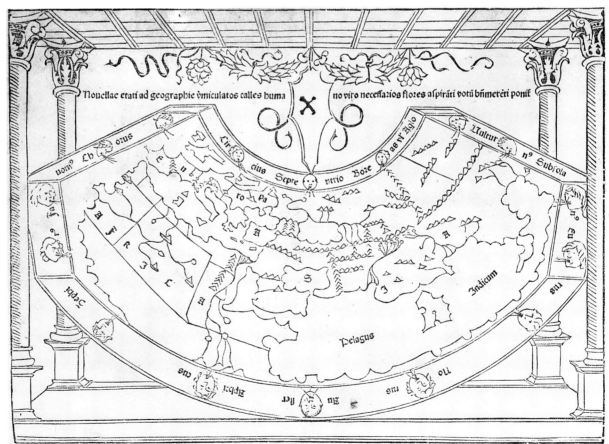

A Ptolemaic world map, printed in 1482, illustrating the works of the Roman geographer Pomponius Mela. The map is set within a Renaissance-style colonnade drawn in perspective: the link between Renaissance art and Ptolemy's geography is made explicit here.

CHAPTER TWO
Maps of the Religious Imagination

FROM THE SECOND TO THE sixth century, the world map is lost in a historic sleep, for it shared with much of western culture the experience of post-classical disintegration. The classical foundations were lost from sight, and when the map re-emerges in the sixth century, it is in a new and quite different form. In this period, the overwhelming fact of Rome's fragmenting secular power is counter-balanced by the growth of Christianity as a civilizing and unifying force in Europe. In the post-classical world, the sources of knowledge and the sources of authority were redefined in Christian terms. The centrality of the Bible had fundamentally altered the approach to all questions of knowledge as compared to the pre-Christian era. Christianity was a faith rooted in history, and the Bible was seen as the record of God's encounter with man, the imprint of the divine on the human world. It consequently assumed an authority in non-religious fields — politics, law, science, art — that was to remain absolute for the next thousand years. In the case of the world map, the authoritative text was considered to be the description in Genesis chapters nine and ten of the division of the world among the three sons of Noah. This was related to the three known continents, and gave rise to the tripartite image of the world which became a cornerstone of medieval geography. Where the basis of the world map in the classical period had been theoretical geometry, it was now the religious imagination.

When the world map re-appears in the post-classical world, it is not in the context of geographical thought at all, but as illustrations in religious or literary works. St.Isidore's *Etymologies,* Orosius' *Seven Books Against the Pagans,* and Beatus' *Commentary on the Apocalypse* and many other works, all contain small, schematic world maps, illustrating either the three continents in the T-O pattern, or the world's five zones, shown as horizontal bands. Elementary as they appear in geographical terms, there are some fundamental and unanswered questions about these early, archetypal world maps. Precisely why certain texts should have been embellished with a world map while others of a similar type should not, is a mystery. It is true that some of these works were concerned with a re-interpretation of the world from a Christian viewpoint, and it might seem natural that a world map should be offered as an *aide-mémoire* to the text. But this does not explain why other works equally concerned with power and culture in the real world (such as Augustine's *City of God,* biographies of Charlemagne; the travels of Marco Polo) should *not* include a world map. More important still is the fact that there is really no conceptual relationship between these maps and their parent texts. The Beatus map for example does not in any sense illustrate the Apocalypse — the other pictures in the Beatus codices do that better. It is therefore difficult to believe that these maps were created anew to illuminate these descriptive texts.

It is more likely that they were adapted from established prototypes and were added to these texts by authors who had a special, if rather arbitrary interest in the image of the world. It may be significant that a number of the seminal maps from this early period — those of Orosius, Isidore, and Beatus — originated in Spain, where Roman customs, the Christian faith and Visigothic law all perpetuated a strongly sub-Roman culture through the post-classical era. It is distinctly possible that the archetypal Isidorean circular world map preserves a now-lost form of late Roman world diagram. The interpretative link with the Bible and with divine government of the world, gave to this form a renewed authority, almost a new dimension. The circle was seen as the natural form of God's created world, while the lines which divided it into three, inevitably recalled the cross.

It is clear that having once appeared with their parent texts, these maps became indissolubly linked with those texts through the mechanics of scribal copying and through the medieval adherence to authority. Almost one thousand extant medieval manuscripts contain small world maps, classifiable into four or five main types, named after parent texts — Isidorean, Orosian and so on. The context remained non-geographical, and the maps developed by a process of literary elaboration, not geographical sophistication. They do not draw nearer to geographical reality between the eighth and the thirteenth centuries, but they become more allusive and enyclopedic. Medieval scholars willingly embraced the corpus or Roman geographical lore bequeathed by Pliny, Solinus, Mela and others, in which the natural and miraculous wonders of the world are of far greater interest than objective geographical description. The project of measuring the earth, the theoretical approach of the classical era, was in total eclipse, replaced by a world map conceived in the imagination. There is no evidence that the medieval mind believed the earth to be flat; there is simply no interest in determining the shape of the earth or of the continents, theoretically or practically. The form of the earth or of the universe was not an intellectual problem, but a religious mystery, enunciated in texts such as Job's: 'He stretcheth out the north over the empty place, and hangeth the earth upon nothing'.

By the thirteenth century the crucial move had been made to free the map physically from its parent text, and large, free-standing world maps, such as the famous Hereford map, were drawn for public display, now transformed into a framework within which to manipulate significant contemporary ideas and images: scenes from the Bible, secular rulers, animals and plants, foreign races and miraculous creatures, the whole subject to God's divine power. The geography remained largely static, since the world of empirical inquiry was closed to the medieval mind, but the world of imagination was not, and within the canons of authority the mapmakers employed cease-

The Garden of Eden from Fra Mauro's world map
of 1459 (page 33). Paradise is located outside the
real world, in contrast to earlier medieval maps, It
is shown as an Italian walled garden, within a setting
of naturalistic hills.

Isidorean Map, 1472, the archetypal T-O miniature.

less ingenuity in elaborating the received map. Imagination led them to develop the symbolic function of the world map, to create an image of the world that was culturally subjective. They neither could nor would create one that was geographically objective. The portrayal of the contemporary world is highly selective: for example the very existence of Islamic power in the Middle East is ignored by the maker of the Hereford *mappa mundi*. These maps, like all forms of learning, were products of ecclesiastical scholarship, and they were naturally dominated by religious landmarks: Rome, Jerusalem, Egypt, Eden. In the twelfth and thirteenth centuries too, there appeared a number of descriptive works entitled *Imago Mundi* — the Image of the World — which were effectively encyclopedias of contemporary knowledge, combining geography with legend, history with fantasy: they were the textual counterparts of the *mappae mundi*.

These maps were complex icons, which evolved within well-defined traditions. Several different types of world map co-existed over many centuries, and there is no indication that any one type was considered to be 'true', or truer than the others. The representation of space was malleable within the demands of the available materials: the exact location of places or symbols was less important than the mere fact of their inclusion. Nor was time fixed within these maps, for the past was always present. Long disused Roman names still appeared on thirteenth-century maps, as did events from the Old Testament, texts from contemporary scholars and images from late Roman fabulists. The great elaborated *mappae mundi* located the events of spiritual history — the Fall, Incarnation, Judgement — alongside London and Paris, Egypt and Greece. The inhabited world of man interlocked with other spheres of existence — that of spiritual history, that of the past, and that of the miraculous, beyond the borders of civilization. The construction of a world map in the middle ages was a literary and a theological exercise, not a geographical one. The medieval scholar showed little awareness of the processes of historical change, and had no inhibitions about introducing anachronisms. On a deeper level of course they were not anachronisms, they were the things which mattered, the texts and the teachings that were authoritative and timeless. In constructing a world map, the essential principle was to adhere to a tradition. This is not to say that the medieval world map remained static and repetitive; on the contrary the extant maps show a multitude of variations. But the structure remains the same, and the variations are elaborations from literary and religious sources, not geographical developments. The map's value lies in its fulfilling its didactic purpose — to display a tableau of the world's forces held in balance: the natural and the miraculous, the human and the divine, the present and the past.

The culture from which these maps sprang was a culture where all learning, even geography, served a central religious purpose, and the map's structure was controlled by the authority of the past. The comparison of this didactic map with the contemporary Islamic world map is deeply revealing. Islam had no ecclesiastical structure comparable with that of Christianity, and Islamic learning was secular. The Idrisi map carried none of the cultural and religious freight of the Christian *mappae mundi*. The medieval Buddhist world map however did share with the Christian map the subjection of the real world to a spiritual landscape. For the Christian, it was the cosmic events of salvation history which dominated the map; for the Buddhist it was the land that had witnessed the Enlightenment of the Buddha that was perceived as the real world. This was not a modified world map designed to illustrate certain themes, since that would imply an alternative, more objective map, which clearly did not exist: this was the only world map for that culture. The *mappa mundi* answered the needs of the Christian ecclesiastical tradition that created it, for it presented a purposeful, interpreted image of the world and of man's place in it. It survived into late medieval art, and even into the era of printing, remaining dominant until challenged in the fifteenth century by secular forces, which would fashion a wholly different world map.

The orb, an enduring symbol of royal power, repeats the medieval T-O world map in three-dimensional form. From a 16th century English manuscript.

Jean Corbichon, 1482, an artistic fantasy based on the world map.

Beatus Map,
1109

ALTHOUGH THE MOST CELEBRATED MEDIEVAL maps are the large free-standing examples such as the Hereford Mappa Mundi, these were exceptional creations. Far more numerous were the small world maps drawn to illustrate religious or descriptive writings. There were perhaps four or five distinctive types of map, which were copied, modified and elaborated in succeeding editions of their parent texts, so that recognizable genres or families of world maps co-existed in medieval Europe. One of the most distinctive of these groups are the maps drawn to illustrate a commentary on the Revelation of St. John written by a Spanish Benedictine monk, Beatus of Liebana. The original manuscript with the prototype map was completed in the year 776, and this copy, dated 1109, still demonstrates the characteristics of its origin in Mozarabic Spain. The entire Beatus manuscript illustrates vividly the visionary poetry of the Apocalypse. The strongly spatial element in the work's imagery — angels moving among the dissolved boundaries of earth and heaven — seemed perhaps to demand a world map. There seems to have been no attempt to map the specific setting of the work among the churches of the eastern Mediterranean; cities such as Ephesus and Smyrna are not named. The map is drawn on the prevalent medieval pattern: east at the top with a vignette of Eden, the Mediterranean in the centre, and islands such as Britain and the Fortunate Isles detached at the base. The Beatus map however is distinguished from other medieval map types by two important features. First, there is the map's rectangular rather than oval shape, undoubtedly derived from the text at Revelation 7:1 'I saw four angels standing on the four corners of the earth'. And, secondly, there is the inclusion of a fourth continent, clearly visible beyond the Red Sea. This conflicts strikingly with the orthodox religious picture of a tripartite world, but speculation concerning a fourth continent beyond the southern sea, beyond the torrid zone and perhaps inhabited by Antipodeans has a long pedigree in classical literature, and is found explictly in late Roman geographical writers such as Pliny, Solinus and Mela. The idea was echoed by Isidore of Seville (c.560–636) whose encyclopaedic works were of seminal influence throughout early Christendom. Although it was apparently a rather abstract and harmless speculation, it carried dangerous implications, for since all men were descended from Adam, where could the inhabitants of the detached Antipodean continent have come from ? This awkward problem explains why the fourth continent remained an exceptional feature on just a few medieval maps. The inscription on this copy states that the land is a desert, unexplored and torrid.

In style the map is both spare and vivid. No attempt is made to fill every space with text or image, while the mountains, the fish-crowded rivers, and the vignette of Eden are drawn with a sinuous energy that suggests a non-European origin: in eighth-century Spain, Islamic motifs and colours were applied to Christian subjects to create the distinctive Mozarabic school of decorative art.

Although clearly a schematic map with a limited purpose, its geographical sources are highly revealing. The regional names — Judea, Bithinia, Panonia etc. — are those of the late Roman provinces, which by the twelfth century had long ceased to have any currency or significance. Ravenna is also named, an echo of its sixth century importance. The image of the world in the Beatus manuscript tradition is comparable to a Byzantine icon, strict in its method, limited in its materials, and having a symbolic not a realistic function.

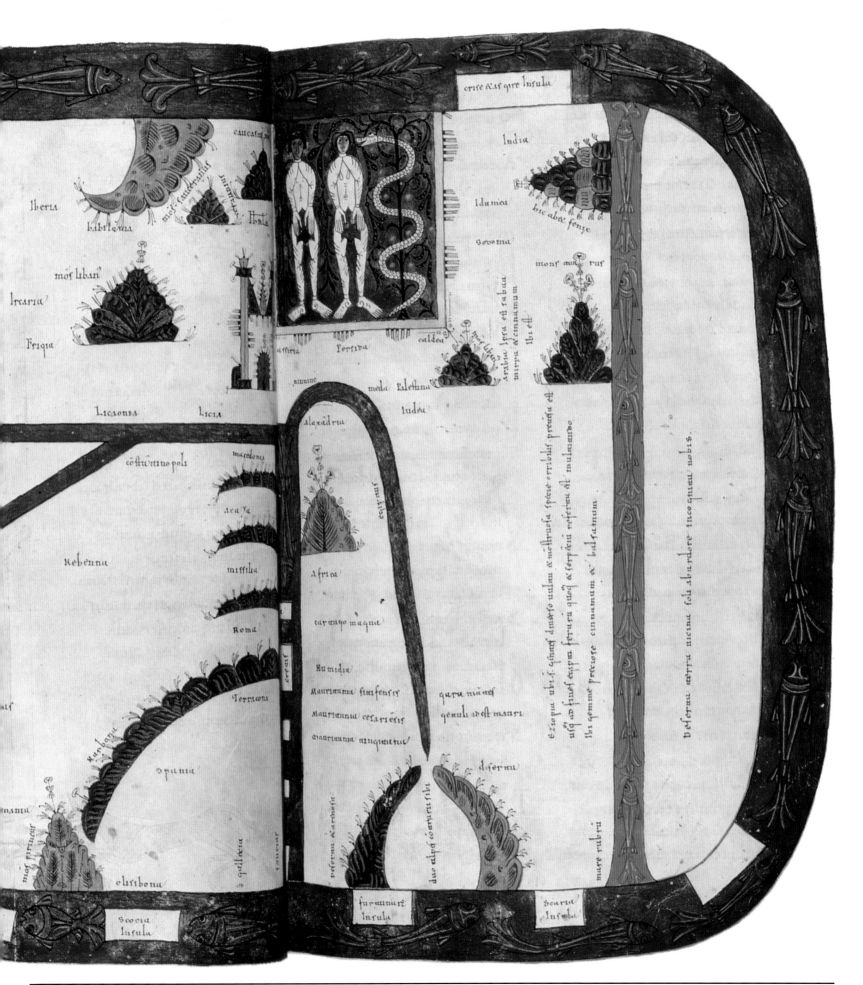

crise &as qire Insula

Iberia
Caucasus
mos emdcranus
mmricorii
Thilia
India
Idumea
Sodoma
bibilrema
mos libani
mons auri rus
Ircaria
Frigia
arabie ipsa est sabia
mirra & cinnamum
Ibi est
Assiria
Persida
caldea
mos libani
Sinaiue
meda
Palestina
Lidonia
Licia
Iudea
alaxadria
cipriis
costantinopoli
macedonia
Acaya
Ravenna
missilia
africa
Roma
cartago magna
Humidia
Mauritania sintfensis
qaru mantes
Mauritania cesariensis
gesuli ad est mauri
Terraconi
Mauritania tingitania
aspania
deferta
refertur & cartania
duo elips eo cetirsi sibi
Spania
mos pireneu
gallicia
olisibona
mare rubru

Eziopia ubi & cinas denario uidias & meonstruosa spene orribile prcusca est usq co fines egiptu ferunt quoq & serpentu refertur & mulacado
Ibi gemme preciose cinnamum & balsamum

Deserta terra uicina soli abardore incognitu nobis

Scoria
Insula
furmurari
Insula
Scaria
Insula

Psalter Map,
c.1250

In STUDYING AND INTERPRETING MEDIEVAL world maps, one of the greatest methodological problems is to work out the relationships among the numerous surviving examples. We must assume that the earliest copies of works like Isidore's encyclopedia and Beatus' commentary on the Apocalypse, both from the seventh century, contained a world map, but our extant copies of these works date from the twelfth century. It is thus almost impossible to reconstruct the intellectual history of these maps during the long centuries of silence. One of the key processes that is largely hidden from us is the evolution of the elementary Isidore T-O map into the elaborate visual encyclopedias of the thirteenth century.

The Psalter map, accompanying a thirteenth century copy of the Book of Psalms, is a key survival in that transition. Although tiny (6" × 4") it exhibits a wealth of detail and many important motifs of the elaborated *mappae mundi*. It is among the earliest maps to place Jerusalem firmly at the centre of the world; among the earliest to symbolize Christ's power as overseer of the world; among the earliest to depict visually Biblical events such as Noah's ark, the crossing of the Red Sea, and the walls imprisoning Gog and Magog; and among the earliest to display the monstrous races in Africa, although these creatures had appeared earlier still in medieval art and architecture. On the evidence of the Psalter maps, all these hallmarks of the mature *mappae mundi* were well established by the early thirteenth century, suggesting a natural connection with the enriched visual art of the twelfth-century Renaissance in France and England. It is recorded that in 1236 Henry III commanded *mappae mundi* to be painted as murals on the walls of Westminster Palace and Winchester Great Hall, and it has been suggested that the Psalter Map may be a miniature copy of the Westminster map. As wall decorations these must certainly have been large and visually elaborated, and the Psalter Map belongs to that period. The importance of the artistic enrichment is twofold: first it transformed the schematic circular disk into the visual encyclopedia we know from the Hereford map; second it developed the theological dimension of the image of the world, clearly visible in the figure of Christ both dominating the world and symbolically holding in his hand a small T-O globe. The image of Christ Pantocrater (Ruler of the Whole World) had emerged in Byzantine art as early as the tenth century, but in western art this figure seems only to have flourished in the context of the Last Judgement. Its appearance here without the imagery of the Last Judgement is unusual in western art.

What is most intriguing about this visual enrichment is that it did not take the form of geographical elaboration. It might be expected that the crusades of the twelfth century would bring a new geographical awareness into European society, as tens of thousands of west Europeans crossed and re-crossed half the known world. But the crusades apparently had no such effect, and the world map, as a map, remained static. As a cultural image, the map was to develop under the guiding forces of theology, iconography and literature, but the encounter with geographical reality was not yet possible. Until empirically-derived coastal outlines and measured spatial relationships were available for any of the regions of Europe or the Mediterranean, they clearly could not be incorporated into the world map, and even when such data was available the makers of the religious *mappae mundi* chose to ignore it. Instead the medieval mapmakers selected symbols of man's history and environment, symbols mainly drawn from the Bible, and arranged them into a coherent image of the world.

The Hereford Mappa Mundi, c.1300

THE HEREFORD MAPPA MUNDI HAS been preserved in Hereford Cathedral since its creation around the year 1300. It is the largest, most detailed and most perfectly preserved medieval map in the world, and its survival is of the greatest importance for our understanding of medieval beliefs. The map itself is circular, with Jerusalem placed at the centre of the world, and the ocean completely encircling the land. East is at the top of the map, and Christ at the Last Judgement crowns the work. Also at the top, but located within the bounds of this world, is the garden of Eden. Unusually for a medieval artefact, the maker has left his name upon the map: he was Richard de Bello, priest of Haldingham in Lincolnshire, who has been identified from independent documents, though little is known of him. Structurally the map belongs to the medieval group of T-O maps, where the world is divided into three parts: the T within the O is formed by the rivers Don and Nile flowing into the Mediterranean, these waters forming the boundaries of the three continents known to the ancient world — Europe, Asia and Africa. The Mediterranean is the central geographical feature, with Crete and its labyrinth, and Sicily with the flames of Etna instantly recognizable. The British Isles are at the bottom left, while the Red Sea and Arabian Gulf appear like a fang at the upper right. Outside the map's circular border are inscriptions and pictures which shed some light on its intellectual background. This background reaches back to late Roman sources — the provincial names in red are those of the Diocletian empire — while an extended inscription attributes the first survey of the world to Julius Caesar. In the bottom left corner the Emperor is portrayed commissioning three surveyors to 'Go forth into the whole world and report to the senate on all its parts'.

The geographical scope of the map, its world picture, is essentially that of the late Roman period, since in 1300 the known world was still the Roman world. The greater part of Asia and Africa and of course the lands beyond the encircling ocean were still unknown. Almost all the place-names are in Europe and the Mediterranean region, and the map's edges shade into the realm of myth and legend, beasts and savages. The biblical lands are given a relatively large area, reflecting their importance to the medieval mind, and many biblical scenes and cities are shown. The peripheral vignettes — the sphinx, the mandrake, the pelican, the cannibals — are perhaps the map's most striking visual feature, so sharply-drawn, so grotesque and childlike. They represent a stratum of medieval lore and belief which appeared repeatedly in literature and art. On another level of reality is the abundance of familiar towns with their towers and city walls, often quite small, as with Edinburgh, Oxford, Salzburg or Cordova, but showing the major cities of Paris, Rome, Antioch or Alexandria with elaborate, terraced architecture. The number and reasonably accurate location of these places is evidence of medieval awareness of real geography.

Here are the two apparently disparate elements of the *mappae mundi*: a detailed reference map of medieval Europe, and an encyclopaedic chart. It is in the co-existence of these elements that the map reveals its true nature: it is no less than an intellectual world-picture, where history and theology are projected onto an image of the physical world. The great cosmic events of past and future — the fall of man, the crucifixion, the apocalypse — are located in the real, inhabited world, alongside London and Paris, Spain and Egypt. And then at the borders of the known world are the reminders of the savage and demonic forces in the created order. Around the borders of the map appear the letters M-O-R-S, recalling that all life is under the shadow of death. The parallel that comes clearly to mind is the work of Dante — exactly contemporary with Richard de Bello. From a real forest on a specific day, Dante entered another dimension, then emerged onto a mountainside which is both a place of the spirit and located in real space. Clearly we cannot now be certain whether Dante or Richard de Bello believed that hell or purgatory were really located in a certain place; but they did believe in their co-existence with the normal world, and in what theologians term salvation-history or spiritual history, both in and out of time, and capable of being mapped in space.

(Reproduced from a 19th-century facsimile.)

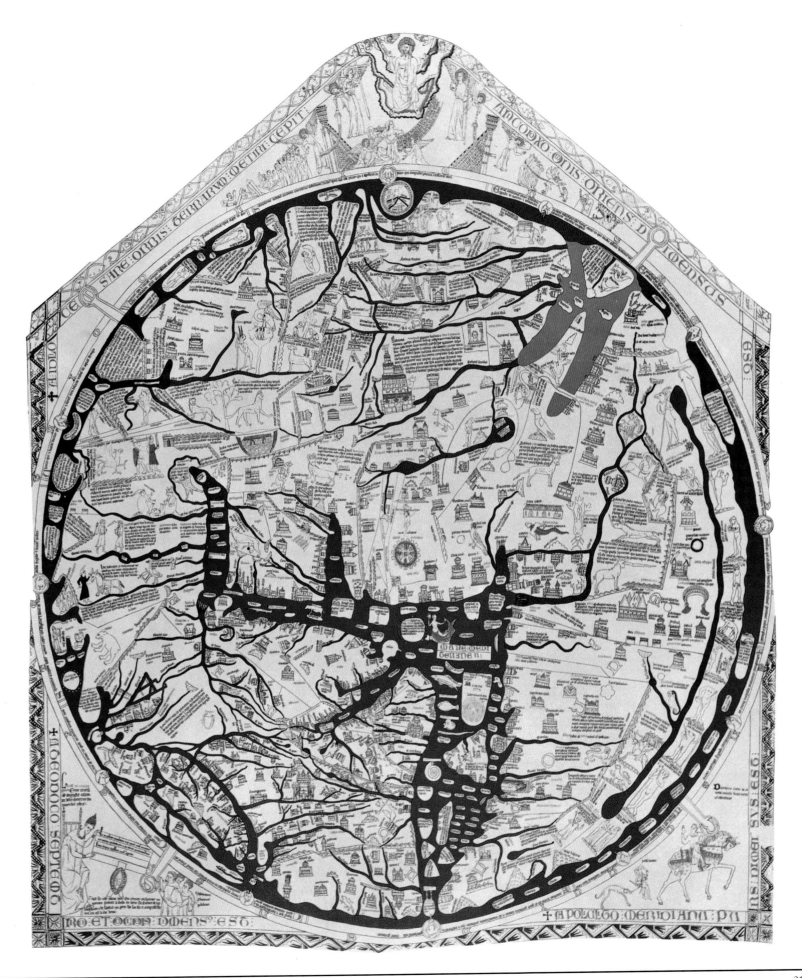

HEREFORD MAPPA MUNDI, 13TH CENTURY

Gotenjiku-Zu, 1364
Buddhist Map from Japan

BUDDHISM HAS EVOLVED AN ELABORATE cosmology which provides a systematic visualization of states of being, and of the attributes and environment of the spirit and the self. On a more concrete level however, Buddhist teaching sought to give a picture of the actual world of man's present existence. This map, a product of Japanese Buddhism, gave expression to those ideas in a form which flourished from c.1000 A.D. to the seventeenth century. In this picture, the central feature of the world is the great mountain Sumeru, and the inhabited world of man (in Sanskrit *Jambu-Dvipa)* lies to its south. From the sacred lake Anavatapta in the north, flow a number of great rivers. It is noticeable that north is the sacred quarter in Indian religions, and in Buddhist thought as in Hindu, these symbolic features came naturally to be associated with specific places in the Himalayas. In particular Mount Kailas in Tibet came to be identified with Sumeru, while the nearby Lake Manasarowar is indeed very close to the sources of the Indus, the Brahmaputra, the Sutlej, and the Ganges. Thus this map presents an Indo-centric view of the world, despite its Japanese origin. The circular lake can be clearly seen towards the top of the map, and the representation of mountains, forests and water is more natural than anything in contemporary western mapmaking.

Other countries than India are named on the periphery of the oval map: China to the east, Persia to the west, and the islands of Ceylon and Japan barely discernible in the south-west and north-east respectively. These place-names and all those on the map are taken from a seventh-century text by the Chinese Buddhist pilgrim Hsuang-Tsang, whose book 'A Record of the Regions to the West of China' is an important historical and topographical source. His route is the red line marked on the map: from the Silk Road he turned south-east through Kabul and Kashmir, and sailed down the Ganges visiting the places associated with the Buddha. In north-west India he saw hundreds of temples and monasteries pillaged by the Huns. He journeyed south as far as Madras, then returned to recross the Hindu Kush, observing, studying and recording throughout his fifteen years in India. On his return to China, a group of disciples gathered around him, including some Japanese who carried his knowledge back to

Japan. Hsuang-Tsang's geography was imposed upon the schematic framework of the *Jambu-Dvipa* map to produce a conceptual world map satisfying to the Buddhist mind because of its concentration on India.

It is remarkable that this map is such a near-contemporary with the Hereford Mappa Mundi, for in both maps we see the religious impulse shaping a symbolic world picture. In the Gotenjiku-Zu the symbolism takes an extreme form, where one region, because of its religious significance, comes to dominate the world map, so that Japan itself is scarcely represented. This divorce from real geography is even more alien to the modern mind than that of the Hereford map, and it raises the question of the level on which this world map was to be understood. The theme of the map is plainly that of pilgrimage to the Buddha's native land. In an inscription on the map, the author has left his personal testimony: 'With prayer in my heart for the rise of Buddhism in posterity, I engaged myself in the work of making this copy, wiping my eyes which are dim with age, and feeling as if I myself were travelling through India'. This makes it clear that the purpose of the map, like the purpose of the Hereford map, was devotional, scholarly, and didactic: it employed the idiom of geography, but it was not geographic. To ask 'Was it a serious world map?' is to miss the point: it was the only world map. For the Buddhist monk, as for the creator of the Hereford map, it was neither possible nor desirable to step outside his culture and produce a different map. These maps located in space and time the history of Enlightenment or Salvation — the heart of reality. Accidents of geography, such as the location of Britain or Japan, existed on a completely different level. No other map was devised in medieval Japan. The Gotenjiku-Zu flourished until the arrival of a totally new world-picture with the European missionaries in the sixteenth century, and even then the traditional map survived alongside the western model. As late as the eighteenth century during Japan's self-imposed isolation, the Gotenjiku-Zu was fulfilling its role as a map of the devotional realm, and Hsuang-Tsang's journey was as real as the imperfect reports of lands and peoples beyond the now impassable sea.

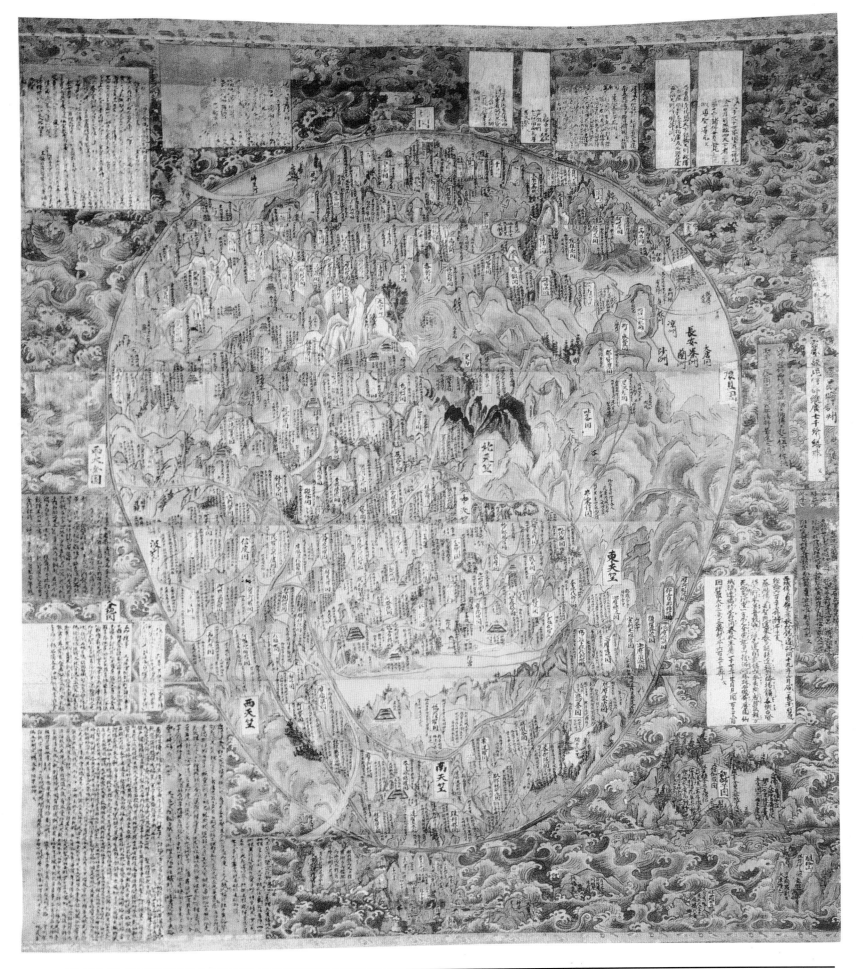

Evesham Mappa Mundi, c.1390

THE LARGE CIRCULAR OR OVAL *mappa mundi* became in the high middle ages a peculiarly English phenomenon. All the surviving large elaborated mappae mundi and many important smaller ones drawn between c.1200 and c.1400 are English or have English connections, while Italy appears scarcely to have shared in this tradition. Why, in the international culture of medieval Christendom, this cartographic form should be virtually a national possession has never been explained. The Evesham map, less elaborate than the Hereford map and drawn about a century later, shows the ways in which the English tradition developed, progressed in some ways, in other ways regressed.

The map was drawn in Evesham Abbey, Worcestershire, commissioned by the man who was Prior there from 1352 to 1392, Nicholas Herford. Oval in shape rather than circular, the map's geographic framework is very similar to the Hereford map. The most obvious difference is the absence of the vignettes of monsters and marvels, and the absence of any portrayal of Christ, indeed only three religious events are referred to: Eden, Babel and the crossing of the Red Sea. The concentration of interest in this map lies in its depiction of England and France, where the overriding principle is English nationalism. Although not drawn with any geographical accuracy — Scotland and Wales are both separated from England by sea — England is swollen to an enormous size. France is diminished, although Calais and St.Denis are marked by flourishing towers. The context of this distorted image is clearly that of the Hundred Years' War. Calais had been the English bridgehead in France since 1347, and the royal abbey of St.Denis was the place of burial of the French kings, and the English kings in this period laid claim to the throne of France. There can be no doubt that this section of the map had the political purpose of proclaiming English power in France. Nor does the English emphasis of the map end there, for it is also a detailed depiction of the region around Evesham itself. More than fifty place-names are given (in vernacular form, not Latin) including Gloucester, Tewkesbury, Northleach and Winchcombe. A further local element is that the vignette of Adam and Eve has, in an allegory of church power, been drawn on the back of a carved throne, which strongly resembles the Abbot's chair from Evesham Abbey (still preserved in Evesham).

The structure of the Evesham map, its world-view, is still the conventional, schematic late medieval structure familiar from earlier maps. But comparison with the Hereford map reveals how a map's character was capable of being shaped and altered for a particular purpose. Compared with the Hereford map, the Evesham map shows a retreat from the visual-encyclopedia concept. With its comparative lack of religious imagery and its political dimension, it is almost a secular *mappa mundi*. In its depiction of England and its gazetteer of place-names, it is in some ways more seriously geographical, although the sense of spatial relationships has still not emerged. With these national and local concerns, one might almost describe the Evesham map as a map of England with a world map added to it. This radical treatment of the world map from within the ecclesiastical establishment, only a century after the Hereford map, raises sharply the questions of tradition versus innovation, authority versus originality in the medieval mind. It is often claimed that medieval thought could never step out of its slavish adherence to precedent. But the reality was that, although a framework of authority was always the essential starting-point, within that framework endless ingenuity could be employed in the expression of artistic or literary themes. In this case we see the world map being re-shaped not by geographical knowledge but by politics. Although it is so different from the Hereford map, the Evesham map demonstrates no less clearly that the world map may be the expression of wider, contemporary, non-geographical forces.

Catalan World Map, c.1450

THE SEA-CHARTS, OR PORTOLANS, which are crucial documents in the emergence of modern mapmaking, appeared in the late thirteenth century in the two maritime centres of northern Italy and eastern Spain. By the mid-fifteenth century the accurate, empirically-based cartography of the sea-charts was being incorporated into a new type of world map, distinct from the religious *mappa mundi* and from the classical world map of Ptolemy, which was now available in Latin translation throughout western Europe. The Catalan region, based on the Barcelona-Valencia-Majorca triangle, became in the fourteenth century a commercial and cultural realm where Arab and Jewish elements were active within the Christian culture. Numerous Catalan sea-charts and two Catalan world maps have survived, the earlier the so-called Catalan Atlas of 1375, and this circular map from c. 1450. Both maps display the leading characteristics of the sea-chart: the network of compass lines, and the flags or shields identifying cities and kingdoms, but they were clearly not produced for use at sea. They may be regarded as paradigms of the chart-maker's craft, logical extensions of his vision beyond the Mediterranean to the limits of the known world. Sea-charts of this period were distinctive for their strict empirical basis: they were working documents and showed only what mariners had observed and could rely on. They were functional, and, like any other functional instrument of that time — the sword, the hour-glass, the saddle — they were secular, and intellectually neutral in their purest form. But clearly a world map could not be constructed at this time on these severe and demanding principles, and this is not the sea-chart in its purest form. The anonymous maker of this world map has combined literary sources for some regions of the world, with empirical sources for the Mediterranean region. Thus we see Marco Polo's narratives, already two centuries old, embodied in the depiction of China, and very recent Portuguese exploration of west Africa, naming Cape Verde, first navigated by Dias in 1444. Moreover the religious element is still present in this map, not only in its circular form, but in the little vignette of Paradise — transferred from Asia to east Africa. The strangest geographical feature is the shape of Africa: at the extremity of the Gulf of Guinea, a river or strait connects the Atlantic with the Indian Ocean, while a huge land-mass swells to fill the base of the map. No place-names appear on it, and it is not clear even if it is to be considered part of Africa, or a distinct southern continent. Stylistically, the visual hallmark of the Catalan school is the series of portraits of desert rulers in their tents, some Islamic sultans, others legendary. They are the first western European maps to acknowledge the presence of Islamic power in the Mediterranean region. Comparison with the Al-Idrisi map suggests a possible Arab influence in the drawing of east Africa and the Indian Ocean, significantly different from the Ptolemaic model.

The interest of this map lies in its uncertain, eclectic identity: circular in form, retaining some religious motifs and traces of medieval legend, yet having the navigational apparatus of the sea-chart, and showing certain Arab influences. There is no title, no dedication and no legend to provide a key to interpreting the map's purpose. A map of this complexity, from this transitional period, raises many questions about the level of reality which the mapmaker intended, and the level of belief with which his contemporaries responded to it. Is it conceivable for example that professional mariners would believe in the vast, smooth, arc-shaped southern Africa shown here? Did the scholars of the new humanism give credence to dog-headed kings? Did theologians accept that paradise, removed from the map of Asia after Marco Polo's travels, could seriously be relocated to Ethiopia? Was it conceivable that, once beyond the gates of Europe, the laws of God and nature had no force, and all things were possible? Is it not more likely that this map (and others of the period) displays within itself different levels of reality and representation? The elements within the map range from navigational accuracy in the Mediterranean, through reconstructed overland journeys in Asia, a gallery of exotic royal portraits in Africa, alongside the mythical fragments which were de rigeur for all such maps, but which were now little more than stereotyped images. The circular framework within which all this appears is purely a graphic convention, since the northern, the southern and the eastern edges of the map are unknown and unexplored. The essential unity and seriousness of purpose which we would expect from a world map is absent. The contemporary parallel that comes to mind is the medieval mystery play: religious in origin, but also satirical, improvisatory, flouting the unities of time and place, and functioning on several different levels to present a powerful, dramatic but not a logical coherent picture of the world.

CATALAN MAP, c.1450

27

Al-Idrisi,
1456 (from 12th-century original)

AFTER THE FEROCIOUS CONQUESTS WITH which Islam emerged onto the world stage, in the calm which followed, the caliphate courts became centres of precocious cultural activity. Early medieval Europe had nothing to compare with the scientific and literary achievements of the Abbasid court in Baghdad in the eighth and ninth centuries A.D. Arab geographers translated Ptolemy's *Geography* before the end of the ninth century, compiled their own co-ordinated gazetteers of the world, and were open also to influences from the east, from Persia, India and even China. Against this background it should not be surprising that of all medieval geographers, the one whose work is most complete and coherent, and the one of whom we have most knowledge, is not a western Christian, but an Arab. Al-Idrisi was of a royal family, born in Morocco c.1100 A.D. and travelled extensively in Europe and North Africa. About the year 1140 he entered the service of the Norman King of Sicily, Roger II. Sicily was then a meeting-place of cultures, and Roger's court one of the most diverse and intellectual in Europe. Idrisi's appointed task was to draw up a comprehensive map of the world with a full descriptive commentary, and this he achieved, completing after some fifteen years' research *The Book of Pleasant Journeys to Faraway Lands* commonly known as *The Book of Roger*. At its heart lay a group of seventy regional maps, conceived and drawn separately, but forming when assembled the most detailed world map of its time.

It is clear that Idrisi knew Ptolemy's work and used it as his principle source, supplemented by Arab and Christian travel narratives and by such maps as he could find. Idrisi left a detailed account of the editing of the maps, describing the collation of classical, oral, literary and graphic sources. The area of the world shown on Idrisi's map was precisely that on the Christian *mappae mundi* : Arabia is central, with Europe (except the north), Asia and North Africa shown in considerable detail. The most significant omission from Ptolemy's model is the absence of any discussion of theoretical matters such as projection and co-ordinate positioning. These regional maps were each drawn as rectangular panels, so that the large composite map would also form a rectangle. But this was no more than a graphic convention, implying nothing about Idrisi's conception of the world's shape. That is revealed in his smaller conceptual map of the world which complemented the regional maps. This is circular, oriented to the south, and bears a strong resemblance to the Christian *mappa mundi,* a structure of which Idrisi was certainly aware from his researches. There are few place-names (some manuscripts have none

at all), even Rome and Sicily are not named. It shares with the Christian *mappae mundi* a schematic but recognizable picture of a number of rivers and mountains — the Danube, the Caucasus, the Indus, the Pyrenees. Unlike the *mappae mundi* however, it bears unmistakable Ptolemaic elements, such as the source of the Nile in the mountains of East Africa, and the eastward curve of the African continent. Idrisi differs from Ptolemy in that Africa has no link with South-East Asia, thus leaving open the Indian Ocean. This is almost certainly based on the direct knowledge or lore of Arab sailors in the Indian Ocean, conceivably even on information from China.

So how does this map differ from the Christian *mappae mundi*? How far is it a specifically Islamic map? One feature of the *mappae mundi* that is strikingly absent is the graphic dimension: no cities, no castles, no animals, no legends, no saints, no monsters. This is, on one level, a cultural difference, explicable in terms of the Islamic ambivalence about image-making (although this was by no means a total prohibition as is sometimes claimed). But it is more than a superficial or cultural feature: this world map clearly has no religious dimension, nor is it encyclopedic or symbolic. Taken in conjunction with the regional maps, it is as nearly as possible, purely geographical. In the text accompanying the map, Idrisi recorded Roger's commission to 'Produce a book explaining how the form of the map was arrived at....concerning the condition of the lands and countries and their inhabitants'. This commission is plainly secular in its intention and its scope, comparable to a Domesday, and bearing no resemblance to the icon-like quality of western world maps of this period. The inclusion of Ptolemaic elements reinforces this secular approach to mapmaking. Simple as it appears, this small map is an eclectic document, concentrated from many sources, and it displays a geographical maturity that is not found in contemporary Christian maps. It is not specifically Islamic, in the way that the Hereford map is unmistakably Christian, rather it is international in character, as was the Sicilian court where it was created. Idrisi's achievement was precocious in its time, indeed it was to be the high-point of Islamic cartography: his work was copied, but scarcely developed by other Islamic mapmakers as late as the seventeenth century, when western models of the world map became dominant in Islamic countries.

(*This map was drawn with south at the top: it has been inverted for the sake of clarity.*)

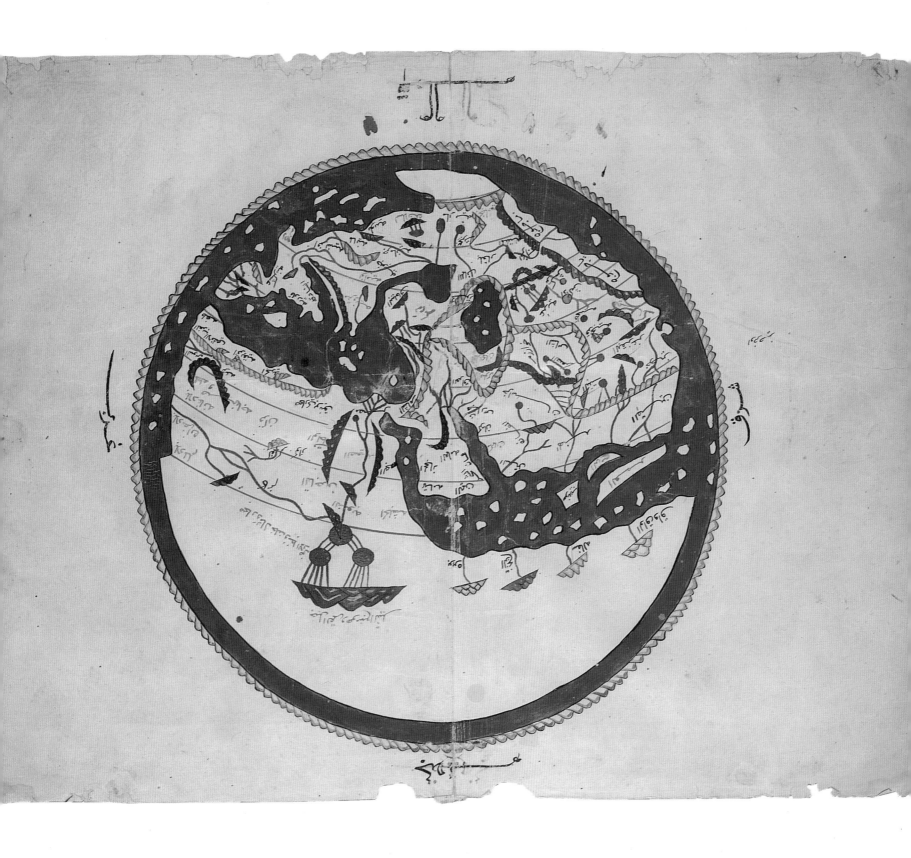

Jain Chart of the World, 15th Century

THE RELIGION OF JAINISM HAS an unbroken history in India reaching back 2,500 years. Although numerically weaker than Hinduism, its adherents have had a deep influence on Indian thought and art. The word Jain derives from the the Sanskrit root *ji* — to conquer, referring to the ascetic struggle through which believers must pass in order to achieve wisdom and freedom. Jainism has a well-developed tradition of figurative charts which indicate the elements of man's earthly and spiritual life. There are many such charts dating from the fourteenth century onwards, and Jain monks would learn how to draw them and expound upon them in accompanying texts.

This example bears the generic title *Manuslyaloka* — the World of Man. It dates from the fifteenth century and its provenance is Rajasthan or Gujarat, where Jainism was most prevalent. The world is shown as three continents, a central disk-shaped land-mass and two outer continents like rings, the three being separated by two circular oceans. The jagged outer line may represent a mountain range symbolizing the limit of the world of man. The four figures outside this boundary are Jain devotees who have achieved *moksa* — liberation. The map is centred on Mount Meru (also sacred to Hindus), upon which four other mountains formed like elephant tusks converge, and from which two great rivers flow. The crescent-shaped land below Mount Meru may represent the Indian subcontinent, punctuated by lakes and rivers.

This is however uncertain in view of the map's outstanding characteristic — its precise symmetry. This symmetry relates to both major axes, horizontal and vertical, so that every feature in each quadrant is exactly mirrored in all three remaining quadrants. This deliberate symmetry makes it clear that, although the chart makes use of some features from real geography, such as rivers and mountains, the structure of the whole is not intended to be representational. The outer circular seas and continents, the latter containing real people, clearly indicate a general belief that beyond India's coasts are other lands and other peoples. This may have been pure speculation, or it may have been based on seaborne contacts with other cultures of South Asia and the Middle East.

In the light of this complete absence of a basis in real geography, the strong tradition of chartmaking in Jainism may appear paradoxical: what was the role of these charts, and why did the religious imagination seek to express itself in spatial imagery? In a general sense they are a further expression of the impulse to visualize the fundamental elements of man's environment, which is common to nearly all cultures. But more specifically the symmetry of the map and its repetitive pattern suggest that it was an aid to contemplation. The religious goal of Jainism is the purification of·the soul achieved through non-attachment to things, to people, and to places. The map's circular structure must derive from the cyclical vision of time and human destiny that is so characteristic of Indian religions. A concern for real locations and accurate geographical structures in a map would be out of place in such a world-view. This world has its own pattern — of land, sea and mountain — endlessly repeated, but without ultimate significance. The symmetry symbolizes the finite, worthless level of existence which must be transcended. Thus the non-representational map becomes a religious icon, and the image of the world becomes unworldly in its purpose. Comparison of this map with the near-contemporary western *mappae mundi* throws into sharp relief the western preoccupation with information, with authority, with cultural accumulation and plenitude. This Indian map mirrors just as certainly the religious psyche from which it sprang, but this time ascetic, contemplative and reductive.

Fra Mauro,
1459

In STRUCTURE AND APPEARANCE THIS map belongs to the school of the medieval *mappae mundi*. Yet its form belies its true character, for it is a map dominated by real contemporary geography. It was created within the ecclesiastical tradition, yet clearly demonstrates the process of secularization that was at work in the fifteenth-century world map. Fra Mauro was a monk in the Camaldolese monastery on the Venetian island of Murano. He was famous among his contemporaries as a mapmaker, although this is his only surviving work, and he was commissioned by King Alfonso V of Portugal to draw this world map. One of the many inscriptions on the map records that the king supplied Fra Mauro with the latest Portuguese sea-charts as source material. The most immediately striking — and disorientating — feature of the map is that south is at the top of the map. Although it was standard among Islamic maps, no other western *mappae mundi* use this orientation. Fra Mauro offers no explanation for choosing it, but we must assume he copied it from the sea-charts with which he was familiar. When once this initial strangeness is overcome, even a brief reading of the map makes it clear that we have entered a different world from that of the Hereford map. Gone are the images from religious history and from pagan mythology; gone are the grotesque creatures inhabiting the edges of the world, and gone is the Last Judgement that awaited the world. Jerusalem is no longer the centre of the world, the Atlantic and Indian Oceans are real seas, not merely a conceptual ring of water, and the Mediterannean coasts are drawn with startling accuracy.

Fra Mauro refers explicitly to his major sources, classical and contemporary; the Ptolemaic model, he says, is not really suitable for a modern world map because our knowledge of the world now extends well beyond the Ptolemaic framework. The single most important source of that new world perspective was the narrative of Marco Polo, from which Fra Mauro quotes freely. So influential was Marco Polo's description of China that it became one of the principal spurs to European exploration in the fifteenth century. As with the near-contemporary Genoese world map, the depiction of the African coast and the Indian Ocean is highly suggestive. The Ptolemaic concept of a land-locked ocean is explicitly rejected, not yet on empirical grounds, but more as a statement of faith, aspiration or prophecy. There had been a persistent tradition since classical times that a route around Africa was indeed navigable, and it seems likely that Portuguese contacts with Arab traders had strengthened this idea.

There is one way in which the Fra Mauro map differs from all others: its surface style, its visual texture, is a repetitive pattern of text and image, composed in blue and gold. No features dominate but no part of the surface is empty, so that the entire map shimmers with its almost hypnotic pattern. In almost any section of the map, the surface resolves itself into miniature landscapes, composed of a towered city, a nearby river and some trees. These scenes are directly comparable to the distant perspectives glimpsed in the background of fifteenth-century paintings. There are no human figures, no animals, and no symbolic emblems. Any explanation of Fra Mauro's intention in creating this unique graphic surface must be conjectural, but that it was intentional is beyond doubt. Why should he have extended the miniature image of an Italian hillside town until spread over the entire world? Did it represent a vision of the effects of good government throughout the world, a conscious refutation of the darker medieval belief that the edges of the world were inhabited by savages and quasi-humans?

The salient fact about the Fra Mauro map is the contrast, almost contradiction, between its form and content. Within the *mappa mundi* framework the forces of Renaissance science and empirical geography are at work. The future lay with world maps of an entirely different type from this — the extended Ptolemy and the enlarged sea-chart. Yet the graphic conventions of the circular religious *mappa mundi* should not blind us to the highly experimental nature of this map. The parallel that comes to mind is the technical development of contemporary painting: the subjects remain constant — Madonna and child, crucifixion, or annunciation — but the treatment, the experimentation with space and form, springs from new knowledge and a new vision.

Rudimentum Novitiorum, 1475

With the exception of simple T-O diagrams this is the first printed world map. It illustrated a general history of the world, *Rudimentum Novitiorum* (Handbook for Beginners), published in Lübeck in 1475. Neither the author of the book nor the artist who drew this woodcut map has been identified. The application of printing to maps was destined in time to transform the social context in which maps were produced, diffused and studied; but in 1475 that transformation was still some way off, and the form and content of these earliest printed maps were still medieval. This map is actually a T-O map elaborated by the addition of features showing physical geography. The continents follow the T-O pattern, but the Mediterranean is not shown: it must be imagined to occupy the empty space between the two woodblocks from which the map was printed. The place names have been set in small strips of type and inserted into the woodblock. Asia appears to be bisected by the rivers flowing from paradise at the top of the map. An unusual feature of the map is the stylized symbol of a hill or mountain used to represent each country. The positioning of these country symbols is quite eccentric: Flanders appears east of Spain, Iceland is adjacent to Portugal, while Greece is north of Rome. Several mythological emblems are drawn: the phoenix, a devil or cannibal, and the tree of the sun and moon. The image of paradise is puzzling in that the two figures both holding branches of the tree of knowledge appear to be both men, instead of Adam and Eve. Whether they are men, and the image has some definite if arcane significance, or whether it is merely careless draftsmanship, it is impossible to say. The map can scarcely be termed religious, despite the traditional vignette of paradise, a portrait of the Pope, and what appears to be a nun on Mount Carmel.

This map is clearly medieval in character, and it perhaps surprizes us that the greatest technical advance in human history should not immediately have signalled a new intellectual era in mapmaking. But of course these first map printers were merely transferring to the new medium received forms and materials; it could not be otherwise. If a printer in a small German town required a general world map recognizable to his readers, this circular T-O map was clearly still valid. Innovation in world maps lay elsewhere in Europe at this time – in Italy principally, where both historical scholarship and the empirical art of navigation were shaping new forms of the world map. Within seven years of this map's appearance, the first printed edition of Ptolemy to be produced north of the Alps was published in Ulm, and the circular T-O map was then finally consigned to history after a life of almost one thousand years since Isidore of Seville launched it on its protean career.

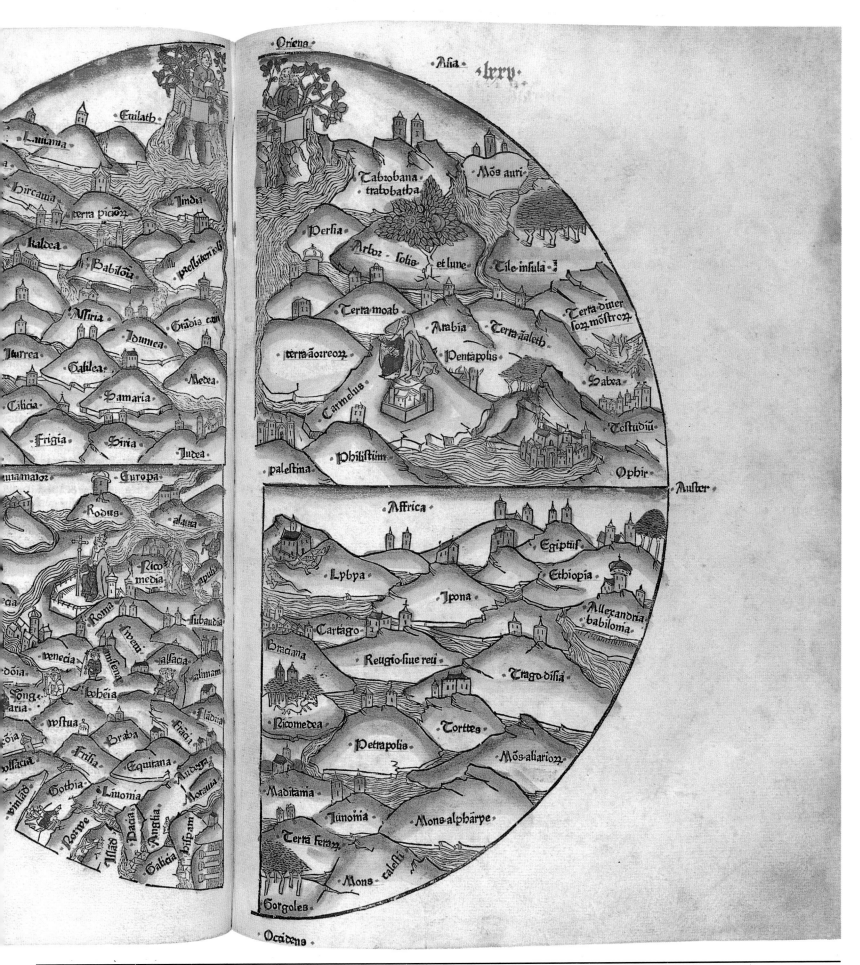

CHAPTER THREE
The Play of Intellect, 1450–1570

'To DISCOVER WHAT LAY BEYOND the Canaries and Cape Bojador; to trade with any Christians who might dwell in the lands beyond; to discover the extent of the Mohammedan dominions; to find a Christian king who would help him to fight the infidel; to spread the Christian faith; to fulfil the predictions of his horoscope which bound him to engage in great and noble conquests and attempt the discovery of things that were hidden from other men'. Thus in 1460 the biographer of Prince Henry of Portugal defined the motives with which his nation sent out a series of maritime expeditions which inaugurated the Age of Discovery. His statement is a remarkable blend of Medieval and Renaissance thinking: religion, legend, destiny, commerce and egotism. In the events that flowed from this manifesto, the years 1450–1570 witnessed the most revolutionary re-shaping of the world map since the end of the classical period. But it would be a mistake to regard that revolution simply as a result of European overseas exploration. Instead there was a whole complex of intellectual, technical and political forces of which the 'Age of Discovery' was itself a symptom, rather than a cause. These forces were in play some years before this manifesto of Portuguese exploration was written, and they had already challenged and partly dissolved the medieval world image.

The first such challenge came from empirical geography in the form of the sea charts drawn in the maritime centres of northern Italy and eastern Spain from c.1300 onwards. These charts were precocious in their accuracy, spare in style and functional in purpose. They were not world maps, being confined to southern European waters, but their clarity and their approach to real geography could not fail to influence the representation of Europe in world maps, and to raise questions about the whole character of the medieval world map. The second revolutionary force was the rediscovery of Ptolemy's geographical works, which re-entered European intellectual life through Latin editions, which appeared first in manuscript in 1406, and in printed editions from 1475 onwards. The concept of measuring the earth, the precise location of places through a system of co-ordinates, the calculated projection of the spherical earth onto a two-dimensional plane, all these came as a revelation to the fifteenth-century mind. In the context of the revival of classical learning, Ptolemy's geography assumed the authority of other classical models in literature, art, and philosophy. It acquired an additional resonance because it coincided exactly with the new awareness of spatial relationships in contemporary art, and the self-imposed task of representing ordered three-dimensional space on paper. Ptolemy provided the ordered space in which a rational world map could be constructed, although as the fifteenth century progressed, its actual depiction of the world became increasingly anachronistic. The overwhelming tendency of these new forces was to secularize the world map, to remove it from the realm of religious iconography. But they did not offer an image of the entire world, and they existed in different spheres: the religious *mappae mundi* survived in the ecclesiastical world, the sea-chart was a tool for trade and travel, while the Ptolemaic map existed in the scholar's study and the princely library. Thus in the mid-fifteenth century, three very different types of map co-existed in western Europe. Only when elements from all three combined could the mature world map emerge. Several important maps from this period — the Catalan, the Genoese and the Fra Mauro — show just such a merging of traditions and a movement towards the secular world map. Secular motivation revealed itself increasingly in contemporary politics and social life. If the middle ages saw human affairs fatalistically, as governed by the blind power of Fortuna, the Renaissance opposed to this the cult of Virtù — the talent, willpower or worth with which the individual could direct his own destiny into paths of his own choosing. Such a cult could be politically channelled into exploration, conquest, the search for wealth and power from

One of the very earliest surviving world globes, made in Nürnberg by Johannes Schöner in the year 1520. The depiction of America and Japan is clearly based on Waldseemüller's 1507 world map.

sources outside the rigidities of European social life. In geographic terms, the external goal, the talisman that drew European ambitions was the East. The influence of Marco Polo's travel narrative had been enormous, creating an image of a distant region whose culture and wealth excited European curiosity and greed. Marco Polo had travelled during a period when the authority of the Great Khan had opened a temporary channel of communication between east and west. Subsequently, Ottoman occupation of the entire middle east became an impenetrable barrier, so that the sea offered the only route out of Europe. Technical developments in ship design, navigation, and armaments were crucial preconditions of this outreach from Europe: the three-masted ship, the compass, the gun, all enabled the world of ideas to interact with the world of experience.

Until about 1490 it proved possible for mapmakers to carry out a piecemeal remodelling of the world map, integrating the new empirical method of coastal mapping with the Ptolemaic framework. There was of course no sudden break with medieval beliefs, and many emblems from geographic tradition were retained, such as the Prester John legend, or the Christian church in India. Even a scholarly and innovative map such as the Martellus map of 1490, which showed the entire African coast for the first time, still includes these symbolic landmarks, although they were later to be relocated to the unexplored regions of the world. But after the Atlantic voyages of the 1490s, the impact of the new discoveries compelled a radical stretching of the world map, and further attempts at the rationalization of geographic space. The problem of projection now became central: when the world map scarcely exceeded the bounds of Europe, projection could be practically ignored. But as the known world appeared almost to explode in its vastness, the problem of displaying it in two-dimensional form became pressing, and in the decades 1490–1520 the Ptolemaic core-map of Eurasia was extended and remodelled until it was no longer recognizable. This was achieved in a series of brilliant improvisations, in which the play of intellect crossed new thresholds, for there was no theoretical science to guide the mapmakers.

Amerigo Vespucci as the surveyor of the New World, from Waldseemüller's 1507 world map, page 48–49.

It is one of the paradoxes of the Renaissance that it was not a scientific movement; it was literary, artistic, political and many other things, but not scientific. Even the Ptolemaic revival was a literary event, a rediscovery of classical theory, whose content was ultimately irrelevant to the fifteenth century. Written works on geography in the fifteenth and sixteenth centuries were indistinguishable from medieval works in their approach. The Nürnberg Chronicle of 1493, or Sebastian Münster's *Cosmographia* of 1544, were still essentially topographic catalogues of an antiquated kind. No theoretical or empirical bases of study were pioneered in this period. In cosmology, the classical, geocentric model of the heavens with its interlocking spheres was still dominant. The first known work to describe and analyse different world map projections was not published until 1590. It was not in geometric science but in the maps themselves that experiment and discovery were taking place. This again parallels the visual arts, where perspective, the new sense of space, was attempted and mastered before it was analyzed scientifically. The Renaissance transformation of artistic forms through the search for reality mirrors the mapmakers' quest for the true shape, extent and image of the world. The play of intellect ran ahead of science and the world map was fashioned into new forms by Rosselli, Waldseemüller, Dürer, Apian and others. The oval world map emerged dominant from this period of experiment, its fundamental intellectual appeal lying in its ability to show 360 degrees of longitude with apparent naturalness. Yet these conceptual developments did not affect the pragmatic, conservative world of the seafarers. Throughout this period, and indeed into the seventeenth century, the sea-chart which mariners continued to use was the unscientific plane-chart model. This type of map was not built on a mathematical projection at all, but simply divided evenly into squares or rectangles of one latitude degree by one longitude degree. The ratio of the sides of the rectangles was constant across the whole map, so no allowance was made for the variable value of the longitude degree; in effect these charts ignored the fact that the earth was a sphere. Such charts were just adequate for traversing small areas or for coasting, but for ocean-crossing they were deeply flawed. They survived by mere custom and practice, because navigators lacked a conceptual alternative. This period also saw the logical extension of the expanded world image: the construction of the first terrestrial globes. However, despite the logical superiority of the globe as a model of the world, it did not displace the map. As an object the two-dimensional paper map was simpler to produce and to study, and it performed also the conceptual trick of presenting the entire world. The globe frustrates because it conceals as much as it reveals: on a world map there is no dark side.

Renaissance science was non-revolutionary because it did not thoroughly redefine the sources of knowledge. This meant that the elements of the world map were not wholly rational or empirical, even by the later sixteenth century. The typical world map title was *Cosmographia Universalis* — a comprehensive picture of the entire world. This was understood by contemporaries as being derived from all available sources, including still the authorities of the past. The moment had not yet arrived in European thought when either experience or reason could outweigh authority as a source of knowledge. All these forces were held in balance by the Renaissance scholar and mapmaker, and the play of intellect had somehow to harmonize them. A rigorous geographical theory that the elements of the world map must be derived only from observation and be capable of being tested, would be alien to the Renaissance cosmographer. The acquisition of knowledge still meant the encounter with the authoritative figures of the past: this was the scholastic approach to knowledge, not yet overthrown by reason or empiricism. Secularization of the map as an object had certainly occurred, which determined both its new form and its new social context. But emancipation from traditional sources, literary and legendary, would have been seen as impoverishing the map. Mercator was the first to employ mathematical principles with the conscious aim of constructing a world map that had a specific property and purpose. But even Mercator regarded himself as a cosmographer in the sense that he consciously presented traditional geographical thought and legend, alongside the recent discoveries of his contemporaries. This explains the endless recurrence on sixteenth-century maps of features such as the southern continent, because Ptolemy or Marco Polo had authorized it.

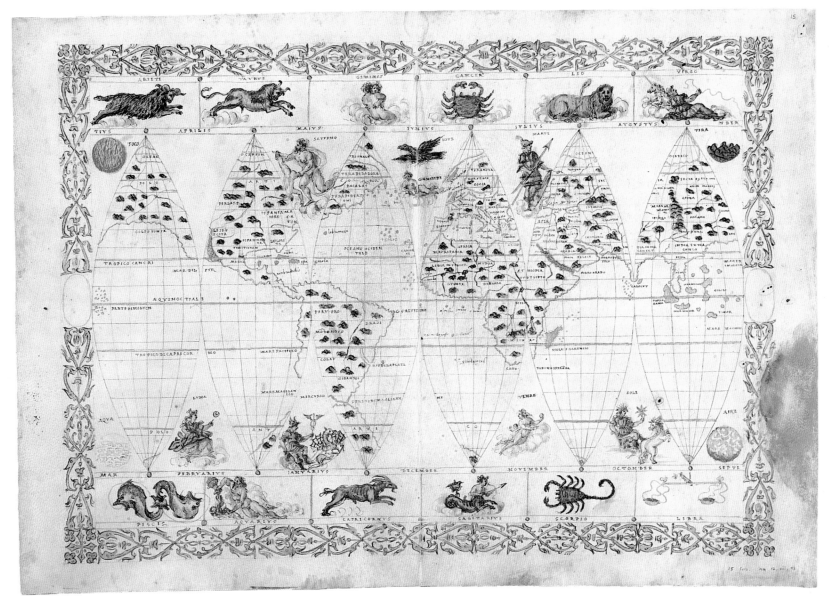

The world map emerged from the Renaissance transformed in structure and intention. Equally important, a new unifying dimension had been added through the medium of printing and the beginnings of commercial map publishing. World maps would continue to display almost infinite variations, but the co-existence of generically different figures of the world had passed away with the manuscript culture of which it was characteristic: printing inevitably crystallized the fundamental concepts of geography. It was also responsible for defining a recognized cartographic language: map elements such as lettering, representation of relief, compasses and lines of bearing, wind directions, engraved lines representing the sea, the title panel — all these had been arbitrary or non-existent in the age of the manuscript map, but now became necessary elements of the mapmaker's idiom. Printing also democratized the map, placing vital geographical information in the hands of any individual or group or nation who wished to use it. The intellectual process of secularization transformed the map of the world from a religious image into a rational structure with both empirical and mathematical elements. The process of democratization that inevitably followed from the technology of printing was to transform it further into an object with explicitly political and commercial dimensions.

A manuscript world map by Francesco Ghisolfi, Florence, c.1565. It is drawn in six segments, as if intended to be assembled into a globe. But the zodiac decorations make it clear that this map was offered as a work of art, and not intended to be cut and mounted as a sphere.

Genoese World Map, 1457

THE INTEREST AND IMPORTANCE OF the Genoese Map lies in its role as perhaps the very earliest world map to show a merging of the three formative traditions of late medieval world mapping: the scholarly, the empirical and the encyclopedic. It is Ptolemaic in shape, designed on a recognizably modern east-west axis, with north at the top, yet its geography clearly derives from sea-charts and travellers' narratives rather than from the Ptolemaic model. The map also retains the character of a visual encyclopedia, derived from the *mappae mundi* showing real or mythical figures all over the world, although the religious theme is now completely replaced by secular history and legend: the emperor of China, warring pygmies, mermaids, Prester John and so on. The map is anonymous and its provenance uncertain: although Genoa is not named, it has always been known as the Genoese Map, and the decorative motifs outside the border show the cross of Genoa and the city wall.

The textual scroll on the extreme left is effectively the map's title: 'This is the true tradition of the cosmographers,' in accordance with that of the sea-charts, from which the frivolous tales have been rejected, 1457'. The map's imagery of Asia is derived naturally from Marco Polo, but also from the narrative of a more recent Venetian traveller, Niccolo dei Conti, who returned in 1444 after twenty-five years journeying in south Asia. The most striking difference from the Ptolemaic world map however is that Africa is shown with a navigable coast, so that the Indian Ocean is not land-locked. Since early in the fifteenth century, Portuguese explorers had been venturing further and further south down the west African coast, with the ultimate objective of entering the Indian Ocean, and so opening a new sea route to Asia, and the Genoese map anticipates this possibility. The picture of a European ship in the Indian Ocean is a striking and prophetic image, as is the brief text 'In this southern sea they navigate without sight of the northern stars', a statement which can only have been based on first-hand reports, possibly from Arab seafarers.

The texts on this map are often confused and unclear, so that a grasp of the map's purpose is tantalizingly out of reach of the modern historian. The text near the west coast of Africa is clearly of great importance in understanding the map: 'Beyond this equinoctal line Ptolemy records an unknown land, and Pomponius Mela and many others, disputing whether a voyage be possible from this place to India, claim that many have passed through these parts from India to Spain'. The reference here is presumably to a south-easterly route into the Indian Ocean, and the overall purpose of the unknown mapmaker seems to have been to present new information about Asia and about the African coast, and to indicate how the latter might hold the key to reaching the former. But a more radical interpretation of the map was clearly possible, not merely to us with historical hindsight, but to contemporaries: one of the most significant features of the map is that its eastern and western edges are bounded by what must logically be the same ocean. The text on the north-western edge of the map reads: 'This sea is called The Ocean, which according to cosmographers stretches out infinitely in every direction, covering the earth except about a fourth part here laid down'. A study of this map would clearly indicate the possibility of reaching Asia by sailing west from Spain.

Henricus Martellus, c.1490

THE DECADES 1480–1510 SAW AN intense, bewildering degree of cartographic development as scholars tried to come to terms with the flow of new knowledge from maritime exploration. This map holds conflicting forces and traditions in temporary reconciliation. The mapmaker was a German scholar active in Italy between 1480 and 1495, responsible for several manuscript editions of Ptolemy and the author of another illustrated geographical book *Insularum Illustratum*. The appearance of the first printed edition of Ptolemy in 1477 did not signal the end of manuscript map production, nor did it fix a standard Ptolemaic world image. Indeed the paradox of the Ptolemaic world map is that it achieved wide dissemination at the precise moment when it was clearly seen to be outdated. This map is Ptolemaic in structure, although it has no latitude and longitude lines. Its outstanding innovation lies in its mapping of Africa, where it draws on the voyages of the Portuguese explorers, culminating in the navigation of the Cape of Good Hope by Bartolomeu Dias in 1487–88. The Cape is clearly seen and named, while the array of place-names extending all down the west coast ends only at the Great Fish River, the furthest point reached by Dias before turning back: the coast thereafter is blank and Madagascar is not shown. The familiar Ptolemaic map of Africa curving east to join South-East Asia is rejected, and Africa itself appears to break through the boundary of the known world.

Comparison with the Ptolemaic map (in the 1477 Rome or 1482 Ulm editions for example) suggests that the Mediterranean and west European coasts have been redrawn by Martellus from contemporary sea-charts: some of the hallmarks of Ptolemy's European coastlines have been corrected, such as the pronounced east-west orientation of Italy, and the eastward curve of Scotland. Scandinavia did not of course appear in Ptolemy's work, and its delineation, together with the placing of Iceland and Greenland were to exercise mapmakers' igenuity until the end of the sixteenth century. Martellus' first peninsula is clearly named for Norway and Sweden, but the second peninsula to the north is an enigma.

In other respects Martellus' map is a rather traditional document. Elements from Marco Polo's account of Asia are sketched in somewhat perfunctorily, such as the city of Quinsai, and the seat of the Great Khan, while the almost obligatory reference to Prester John appears in north India. Breaking the link between South-East Asia and Africa seems to have increased the mapmaker's uncertainty about the whole geography of South Asia — the Indian peninsula, the island of Taprobana (Ceylon), the Ganges mouth, and the Malay peninsula — since Martellus re-locates the Christian church and community of St. Thomas from its traditional setting in south India to southern Malaya. Martellus, as a Renaissance scholar in the Ptolemaic tradition, has no taste for the graphic medieval circus of pygmies, mermaids, giants, wolfmen or cannibals.

We see in Martellus' map a process wholly characteristic of fifteenth-century mapmaking: a merging of traditions, a re-working of classical or received models to accept new features. Where there is modern empirical material, it is used; where there is not, the authority of the Ptolemaic model prevails. In the succeeding decades this process would be repeated, and the lineaments of Ptolemy's map would be re-shaped one by one. The Martellus map has been deservedly celebrated as the last view of the eastern hemisphere alone, the Old World on the eve of the Atlantic discoveries.

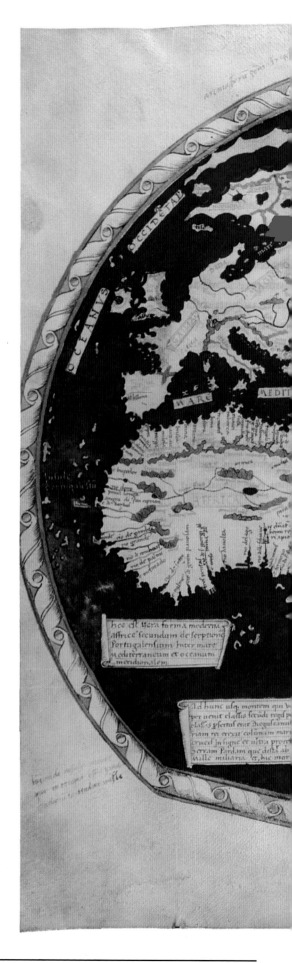

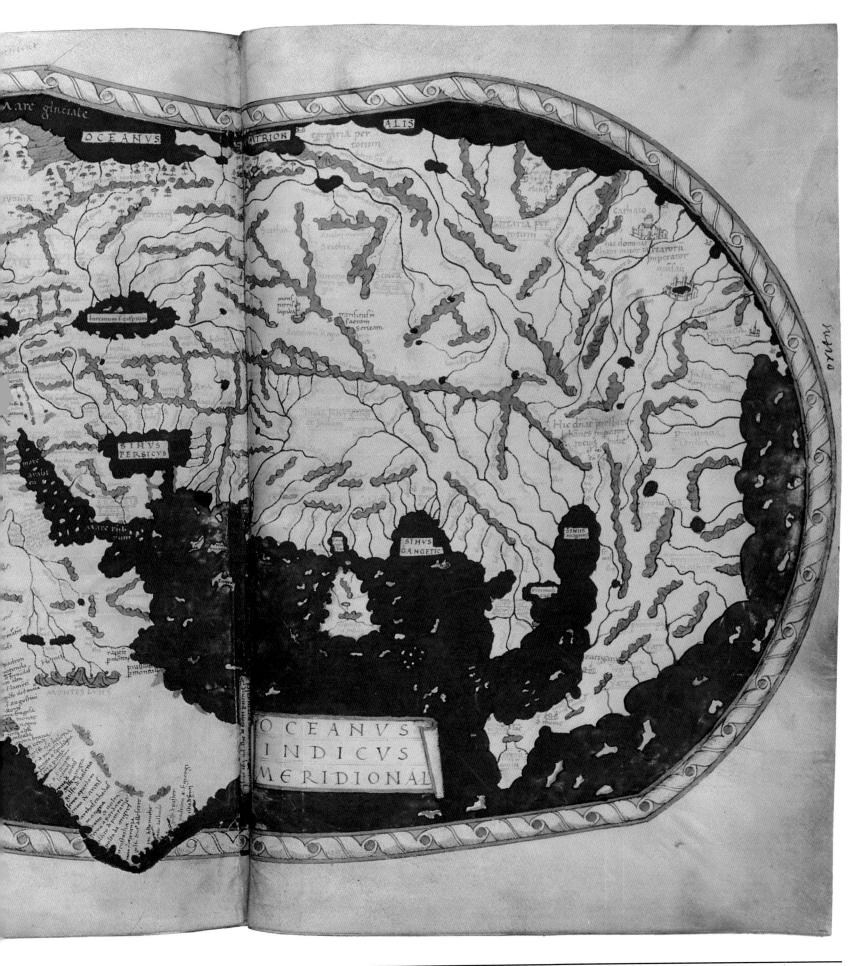

OCEANVS INDICVS MERIDIONAL

Cantino World Chart, 1502

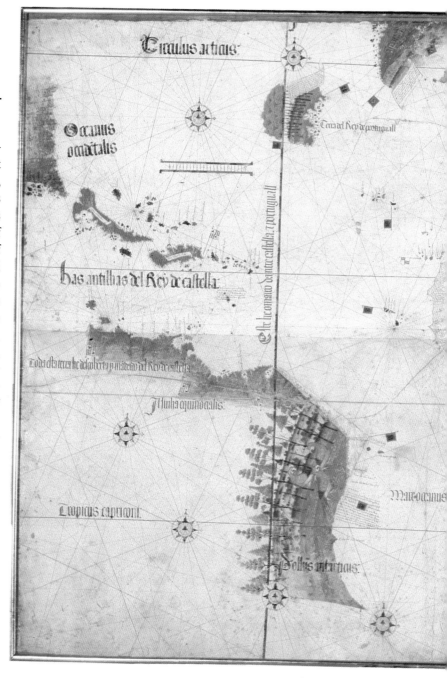

SEVENTY YEARS AFTER THE PORTUGUESE inaugurated the Age of Discovery with their voyages beyond the Fortunate Isles and down the coast of Africa, this map, the oldest surviving Portuguese world map, displays the revolution in geography that had followed. The sea-charts of European and more distant waters that were produced in this period by Portuguese seafarers, were edited into a master-chart of the known world, and this royal chart became a state document of great secrecy and importance. This map was undoubtedly copied from that royal prototype, by an unknown hand, and it is known to have been smuggled out of Lisbon in 1502 by an Italian agent Alberto Cantino, and delivered to his employer, the Duke of Ferrara. It was subsequently seen by Italian and German cartographers, whose own printed maps were deeply influenced by it. The contents of a royal manuscript map like this would normally remain secret for many years or even decades, but here is a rare historical example of a single identifiable document carrying new concepts from one centre of learning, and one country, to another.

One of the first great political questions of the Age of Discovery was the legality of the rival European claims to newly-discovered lands. In 1494 papal arbitration, published in the Treaty of Tordesillas granted to Spain all the lands west of a meridian line lying 960 nautical miles west of the Cape Verde Islands; this line is clearly seen on the Cantino chart. The delineation of Africa is far superior to any earlier map, the coasts punctuated with Portuguese flags and the landmark stone pillars — the padrâos — which the sailors set up at their landing-places. The V-shaped peninsula of India, visited by da Gama in 1498, appeared here for the first time. The depiction of the Americas consists of a large southern land-mass, a smaller enigmatic north-western land, and the Caribbean islands. On the relationship between America and Asia, the chart is ambiguous, leaving open the possibility of a northern link, and also, more surprisingly, referring to Greenland as possibly 'the point of Asia'. In such legends the unknown chartmaker ventures outside the strict canon of the porto-lan chart that everything that is mapped should be empirically based. No literary or legendary material is used to fill, for example, the empty spaces of Asia. The only geographical area where conjecture appears is in South East Asia, unvisited by European mariners until 1511, and here drawn presumably from the Ptolemaic model. With that exception, the map is consciously restrained within empirical limits. Compared with its Spanish counterparts, it is artistically graceful and understated. The whole map is symmetrically balanced around the massive compass rose in the centre of Africa, which is the tangent of two circles each of sixteen smaller roses. This symmetry is slightly disturbed by the Tordesillas line, although the clear sense of the map is that Africa and the sea-route to the east are the great prizes, as envisaged by Prince Henry six decades earlier. The Cantino

chart is a conscious display of Portuguese achievement and political rights in its heroic age.

The context from which the Cantino chart emerged was the elite circle of the court and the court's political and technical advisors. Such maps were associated with wealth and power, and, although clearly anchored in the real world, they also contained esoteric knowledge. The role of the mapmaker in this society lay somewhere between that of the goldsmith and the physician: there was technical skill and mastery of arcane knowledge, but not yet a scientific basis, and the skills were at the service of private patrons. The artefacts which the manuscript mapmaker created were superbly burnished, and they were prized in their time as they are today, but the future lay elsewhere. It lay in the medium of the printed map through which

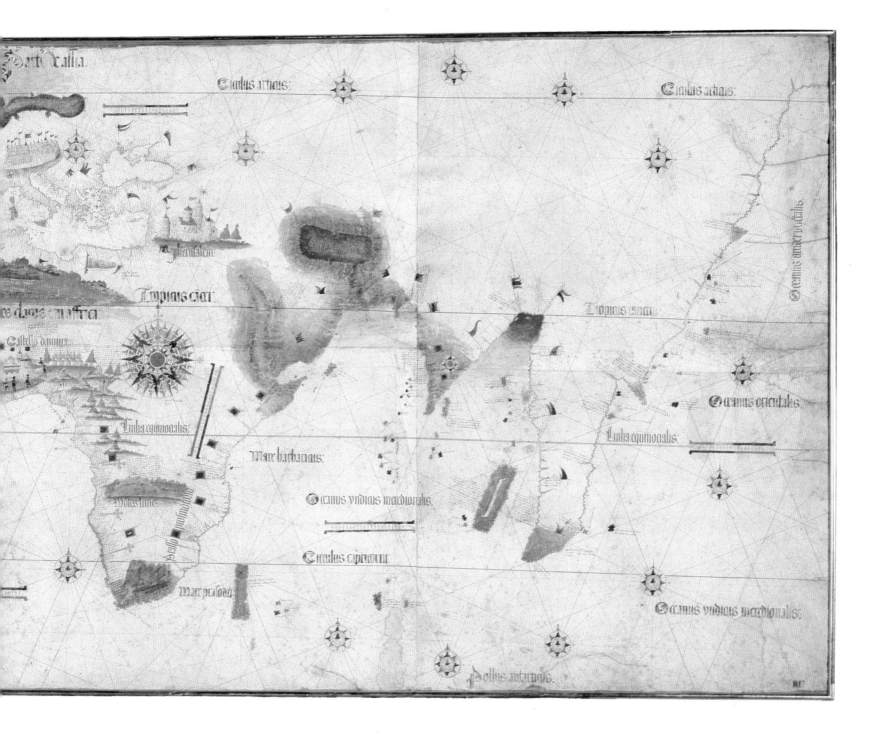

the scholar could exchange and develop geographical ideas in public, and the commercial engraver and publisher could conduct business. Geographical secrecy could not long be secured, and these two worlds co-existed as equals for only the few decades 1480–1530. The interaction between them can be clearly seen by comparing the Cantino chart with the printed world maps by Contarini, Rosselli, and Waldseemüller. After a time, the manuscript world map functions only as a private work of art, while significant geographical progress is recorded in a succession of printed world maps.

Contarini,
1506

Wʜᴇɴ ᴄᴏʟᴜᴍʙᴜs ʀᴇᴛᴜʀɴᴇᴅ ꜰʀᴏᴍ ʜɪs epoch-making first voyage in March 1493, his discoveries were recognised as having decisively shifted the European world-view. But what had he discovered? How was his journey to be understood? After that first voyage, six further transatlantic expeditions were completed in as many years (by Columbus, Cabot, and Vespucci) yet by the end of the century no new map had been published embodying the new world image. Why? First, the sheer uncertainty of where these men had been; second, the rivalry among explorers and their patrons, and their consequent desire for secrecy; third, the intellectual task of remodelling the world map. This is the first published map to display any part of the new world. The mapmaker, Giovanni Contarini, has left his name and the date on the map, but nothing is known of him or the circumstances of its publication. The crucial question among explorers and mapmakers for several decades after 1493 was to determine the relationship between the New World and Asia. Contarini clearly shared Columbus' own conviction that beyond the newly-discovered islands of Hispaniola and Cuba lay the coast of Asia. The land which now appears to us to be North America is clearly named Cathay, together with the features familiar from Marco Polo's description of China. Zipangu — Japan — is equidistant between Cuba and China, while the island of Java is prominent further south. The eastern section of the map displays the received Ptolemaic image of Asia, essentially similar to that in the Martellus map. The huge south-western land-mass coasted by Vespucci and by Columbus on his third voyage, now bears the name *Terra S.Crucis* — Land of the Holy Cross — but its relationship to the continents north and east appears to be non-existent. Here then, thirteen years after Columbus' return, is the first published statement of the new world-view — unambiguous, full of conviction, and totally misconceived.

But the task of interpreting the Contarini map is more complex than identifying new islands and coasts. The most remarkable feature of the map is its fan-like structure, which, although it is not graduated for longitude, appears to indicate a longitudinal extent of 225 degrees. But in that case a massive 135 degrees would remain unmapped between the map's two edges, which is plainly impossible since the coastal strip on the west adjoins the Asian mainland on the east. The answer is that the map is not, as it appears to be, a polar projection, plotted through 225 degrees, with one section unfilled. Instead we must imagine that the world map has been split north-south at a point opposite the viewer, and the two edges have been pulled out like a cloth. The entire world map must be conceived to be shown, but the 360 degree circle has been visually contracted into an apparent 225 degrees. It is as though the world map had been transferred onto a cone placed around the northern half of the globe, and the cone then partly opened. But Contarini has also radically revised the distances involved. The distance from Spain west to 'Cathay' appears to be as great as the eastward distance, whereas the whole Columbus enterprise was driven by the belief that it was approximately one third of the eastward journey. Contarini has in fact, whether consciously or not, drawn what is now termed an interrupted projection: the earth's surface is graphically 'peeled', and the intervening spaces between the segments do not exist as part of the map. But Contarini appears to have squeezed a large segment of that space back into the Atlantic Ocean. The result is an impossibility — an arc of rather more than half a circle which nevertheless represents 360 degrees. Contarini has drawn what much later came to be termed a conic projection, but by omitting the measures of longitude, he obscured what he had done. The Contarini map was copied just once, in an edition of Ptolemy published in Rome in 1507, and with that exception the problematic fan-shaped world map never re-appeared.

The difficulty of trying to rationalize what has taken place in this map illustrates perfectly

the task that confronted the mapmakers of this generation. As in twentieth-century cosmology, space appeared to be stretched and distorted in a way that strained the imagination, especially when the attempt was made to realise the new model in graphic form.

Martin Waldseemüller, 1507

THIS MAP IS OF MONUMENTAL significance in the history of cartography. It summarised the revolution in geography of the preceding twenty years, and it expanded the contemporary image of the world into an essentially new vision. With the advent of map printing c.1475, the Ptolemaic model became the dominant world map, with numerous editions of Ptolemy's *Geography* printed in Italy and Germany between 1477 and 1510. But these were precisely the years during which the Ptolemaic world-view was being exploded by the voyages of Dias, da Gama, Columbus, Cabot, Vespucci and ultimately Magellan. We can only conjecture why, after 1492, there was a fifteen-year wait before mapmakers published new images of the world, but by 1507 Rosselli, Contarini and Waldseemüller had offered the first printed maps showing at least part of the new world. Of these, Waldseemüller's was by far the most radical, indeed in the context of its time, the novelty of this map must have been little short of breathtaking. It was one of the very first separately-printed world maps (i.e. not bound as a page in a book), and it was by far the largest and most detailed map printed to that date. Renaissance manuscript maps were often large, sumptuous and detailed, and it was Waldseemüller's innovation to transfer that grand concept to print and create a map almost 2.5 metres broad, printed from twelve separate woodblocks.

Martin Waldseemüller (1470–1518), active in St.Dié in Lorraine, was a member of the scholarly circle centred on the court of Duke René II. Although neither Waldseemüller's name nor a date appear on the map, its background is known beyond doubt from the geographical treatise — *Cosmographiae Introductio* — which accompanied the map's publication in 1507. From internal evidence it seems almost certain that Waldseemüller had seen and learned from the Cantino chart, whose clandestine removal from Portugal to Italy was so important in spreading the new geography throughout Europe. Nevertheless Waldseemüller's map is constructed on the Ptolemaic projection, and the sections of the map showing Europe, North Africa and western Asia are drawn on the Ptolemaic model. The radical innovation is that the map is stretched from its classical prototype to show 360 degrees of longitude, the first map ever to do so. It is still not a map of the whole world since its north-south extent is curtailed to 130 degrees. The curious bow-shaped top edge of the map is explained by its retention of the Ptolemaic projection up to 65 degrees north, then converging the meridians at the pole, as they must. This gave rise to a depiction of a north coast of Europe and Asia that was wholly theoretical.

This map has a unique historical importance in that it bestowed the name *America* on the new world for the first time. Waldseemüller had read the account of the voyages of the Italian merchant-adventurer Amerigo Vespucci, published under the title *Mundus Novus* in 1503. Columbus himself apparently never abandoned his conviction that the West Indian islands he discovered were islands off the east coast of Asia. Vespucci however divined the truth, that this was neither Cathay nor the Indies, but a new world. Waldseemüller accepted this view and proposed to honour Vespucci by bestowing his name upon the new land. The portraits dominating the map are those of Ptolemy as geographer of the old world, and Vespucci as the guide to the new.

The importance of Waldseemüller's 1507 map rests on a number of points, but it also contains a considerable riddle. America is shown as an island continent, having a mountainous west coast, and an ocean beyond that, stretching to the coast of Asia. Yet Magellan's great Pacific voyage was still fifteen years in the future, Balboa's celebrated sighting of the Pacific was six years away, and all other maps until some in the 1520s, and many after that, betray complete uncertainty about the relationship between America and Asia. Did Waldseemüller

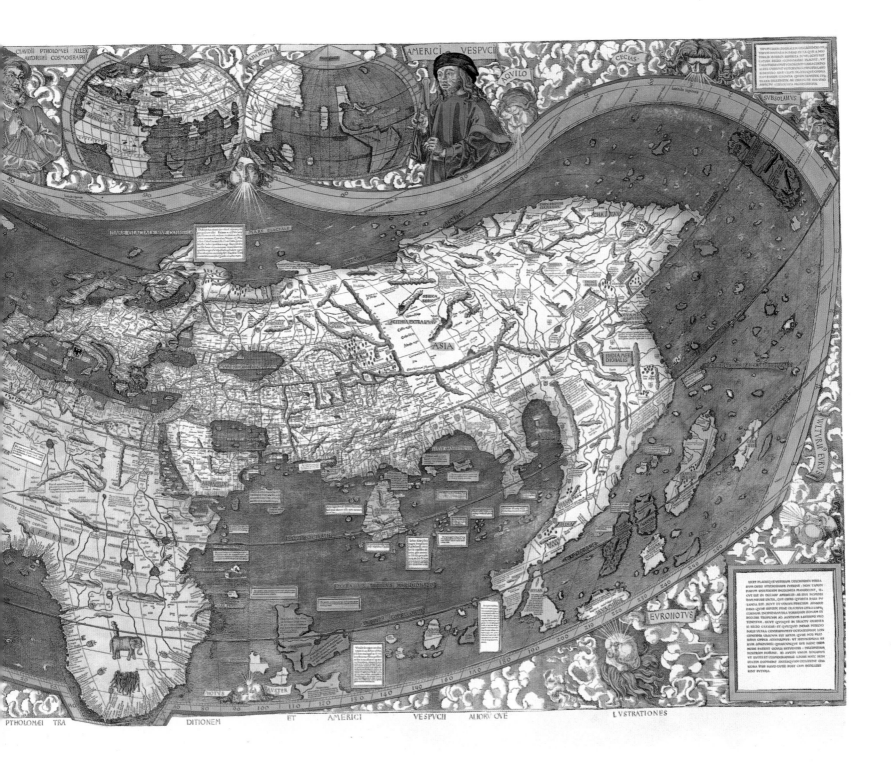

have access to some now lost account of the Pacific, or was this merely an inspired theory? His knowledge of America was certainly not accurate in detail, since he shows a strait dividing North from South America, and he shows no southern passage into the Pacific. The enigma remains unresolved, and adds to the fascination of this great and innovative map.

(*This map has been coloured in style of the period.*)

MARTIN WALDSEEMÜLLER, 1507

Francesco Rosselli, 1508

THIS SMALL, ELEGANT MAP IS, in a very real sense, the first map of the whole world: it shows 360 degrees of longitude and 180 degrees of latitude. Although the degree lines are not numbered, there are 36 lines of longitude and 18 of latitude, so that the clear inference is that each line represents an interval of 10 degrees. This is the first known example of the oval projection which became the dominant form of the world map for much of the sixteenth century, and its visual elegance and intellectual appeal mark a definite turning point in the history of the world map. As an image of the spherical earth, it had a naturalness which made it the most intellectually satisfying model yet devised. It undoubtedly consigned the Ptolemaic model to history, and pushed the plane-chart of the seafarers (see below, the world map of Pierre Desceliers) into its own magnificent cul-de-sac.

We know little of Rosselli's life, but several other important maps from his hand have survived. He came from a Florentine family of artists and engravers, and pursued his career in various European centres before establishing in Florence what is considered to be the earliest commercial workshop where maps were engraved and sold. We have no account of Rosselli's conception of the oval projection, but part of its genius lies in its simplicity. A central meridian and a central parallel are drawn, intersecting to form a cross, their respective lengths having the ratio 1:2. These become axes along which are marked regularly spaced intervals representing 10 degrees. The meridians are drawn as ellipses, the parallels as straight lines. The accurate modelling of the ellipses from pole to pole would have been a difficult feat of draftsmanship.

In common with the other surviving world maps from the early sixteenth century, the geography of America and east Asia is essentially hypothetical. On the one hand Rosselli retains the conviction of Columbus and Cabot that the more northerly landfalls were in parts of Asia, while Vespucci's insight that a new world had been discovered is applied distinctly to South America, here referred to as *Land of the Holy Cross or New World*. North and South America (as we now know them) are guessed to be divided by an extensive sea. On this basis, Asia is massively extended to 280 degrees from Constantinople to Newfoundland, since the great key to this mysterious jigsaw — the vastness of the Pacific — was still hidden. The southern continent, postulated by Ptolemy as a counterweight to the northern land-masses, appears here for the first time as *Antarcticus*. Rosselli's map is unusual in depicting this as a concentrated land-mass, and in locating it south of Africa, but no specific source for this idea has been identified.

The importance of the Rosselli map and its future influence lie undoubtedly in its design and projection rather than its geography. It was the most natural two-dimensional image of the entire world yet devised. Despite its elegance it becomes of course progressively distorted away from the two principal axes. Just how distorted can be gauged by considering for a moment reversing the ratio of the axes, so that the central meridian displays 360 degrees and is twice the length of the equator (*see* illustration, right). The resulting map is startlingly unfamiliar, and apparently useless. But given that the earth is a sphere, there is really no greater validity in the Rosselli map than in the other. Only the central section of the map around the intersection of the axes is reasonably distortion-free. The Rosselli map appears to make sense for two reasons: first the northerly orientation of the map relates logically to compass navigation, universally used by the early sixteenth century; and second, exploration beyond 60 degrees north or south was many years in the future, hence maximum distortion could be concentrated in the polar regions without alienating the viewer. But we are dealing

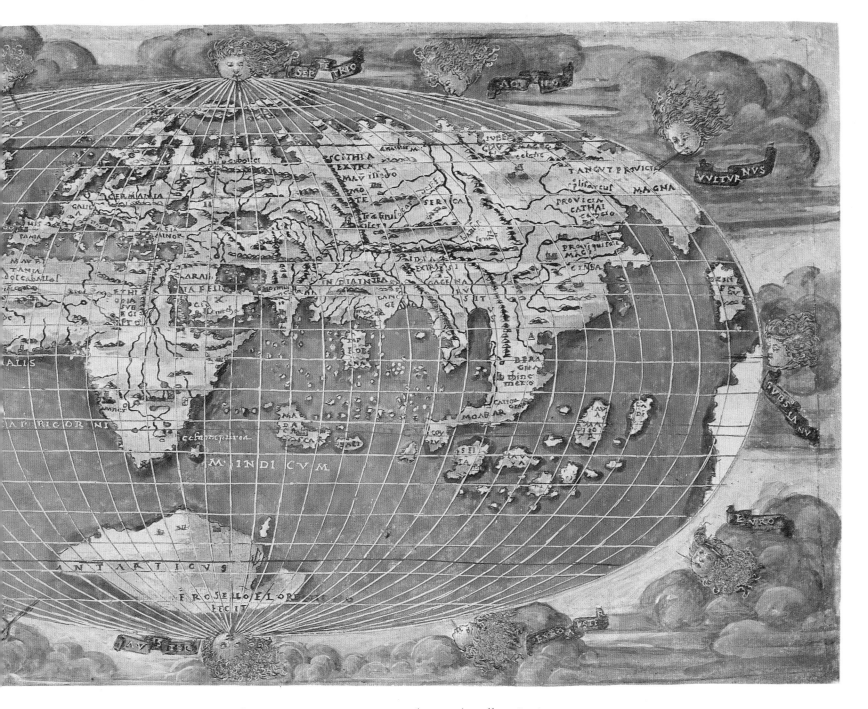

here with optical sensibility and graphic convention, not geometric logic. What all projections seek to do is to give the human eye a god-like perspective of the entire planet, and we need to remind ourselves repeatedly that this is an impossible quest.

Dürer-Stabius, 1515

THAT THE WORLD WAS A sphere was known throughout the middle ages and there is even some evidence that the question of map projection had been perceived as a theoretical problem, by Roger Bacon for example in the *Opus Major* of c.1270. But it had little practical importance since the known world scarcely exceeded the bounds of Europe. It was only when new knowledge enlarged the known world that the problem came to the fore, and solutions were devised which involved stretching Ptolemy's projected map of Europe and Asia in various directions. It is particularly noticeable on all the important maps from 1490 onwards, how prominently the lines of latitude and longitude were drawn: it was almost as important to show how the world was being graphically encompassed as to show the new land-masses. The grid lines must serve to express both the dimensions of the earth and its sphericity.

This measuring, re-ordering and re-designing the world on paper, was clearly related to the work of Renaissance artists who sought to delineate real space in a entirely novel way through the technique of perspective. It is significant that the greatest artists of the southern and the northern Renaissance, Leonardo and Dürer, both experimented with mapmaking. Leonardo toyed with the idea of drawing a world map in segments which might be assembled into a globe. Dürer drew star-charts of the northern and southern heavens, and, in association with the Nürnberg scientist Johann Stabius, drew this experimental world map.

The two texts on the map, a dedication and a publishing privilege, offer little explanation or commentary on the map's purpose, except for the use of the words 'this imaginary orb'. This curious phrase suggests that Dürer and Stabius were aware that they were attempting something quite novel: to show the world as known to the scientific imagination but never seen by man. It is in fact the first map of the world as a real geometric sphere, with the map plotted in visual perspective on it surface. To our eyes it is a prophetic image of the world as seen from space. The similarity of the wind-faces to angels outside the earthly sphere is certainly a deliberate artistic conceit. Did Dürer conceive the idea for this map from studying the earliest globes at Nürnberg, such as the Behaim Globe? It is possible, but the Behaim globe does not show latitude and longitude. Dürer's task here was to plot and extend a modified Ptolemaic map onto the the visible half of the globe. It was the first time such an idea had been conceived, and it prefigures the double-hemisphere map that was still many years in the future. Clearly he was unable at this time to match this orb with a second, showing the western hemisphere, since the Pacific Ocean was still unknown. For the immediate future Dürer's experimental world image had no successors. It was an exercise of that faculty which was to be increasingly important in sixteenth-century maps, the intellectual play, the *jeu d'esprit*. It derived from man's new-found sense of mastery of his world: here the geometry of the sphere is used to create the illusion of seeing the world from a new perspective, and the playful use of angelic faces invites the viewer to participate in this novel experience.

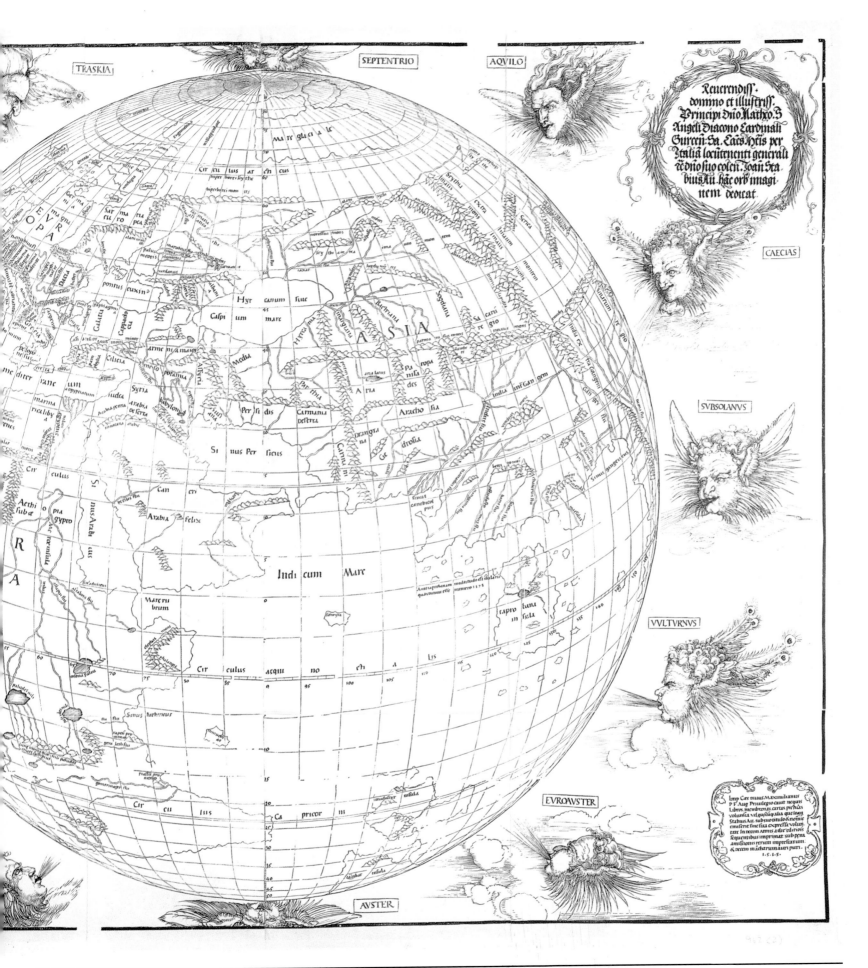

Martin Waldseemüller: Carta Marina, 1516

NOTHING WOULD SEEM TO DEMONSTRATE more clearly the fluidity of the world map during the early years of the sixteenth century than a comparison of this map with Waldseemüller's great map of 1507. Here he has abandoned the Ptolemaic model and, most strikingly, totally recast the shape of America and its relationship to Asia. The polar sea has vanished, as has the Pacific Ocean and Japan, while south Asia is entirely redrawn. What had occurred in less than a decade to cause Waldseemüller to reshape so radically his entire geographical theory? That question may be answered by posing another more precise question: On what level is the formal irreconcilability of these two maps to be understood?

The title to the 1516 map provides the key: *A Portuguese Navigational Sea-chart of the Known Earth and Oceans....* This map is in fact the first and only printed version of the world charts previously known only to Spanish and Portuguese explorers and their patrons. Its direct model is clearly the Cantino chart (see above page 44), and we see again that rare historical moment when a single identifiable document carries new knowledge from one society to another. Waldseemüller's 1516 map is a deliberate expression of that empirical sea-chart tradition which co-existed with the scholarly Ptolemaic legacy. With its plane-chart construction (see below, the world map of Pierre Desceliers), its compass-roses and compass-lines, and its restriction to 250 longitude degrees, Waldseemüller is setting down a record of the known world, empirically constructed from the knowledge of professional mariners. The 1507 map was a theoretical extension of the Ptolemaic model, where literary sources were also accepted. The depiction of Japan is a clear illustration of the difference between the two maps: no western seafarer had visited Japan in 1507, but there was a well-known account of it (as Zipangri) in Marco Polo's work, so Waldseemüller included it. But on the 1516 map it was very properly omitted, since no precise navigational data on it was available.

So we see here Waldseemüller drawing two distinctly different world maps from two geographical traditions, one scholarly and theoretical, the other empirical. He may already have perceived that there would be no single, definitive world map, that intellectual context and graphic convention would inevitably mould any representation of geographical space. But context and convention are what is lost from any work as it moves in time or place: these two maps, seen together, would inevitably provoke the question: which is correct? There is no doubt that the map of the future was the 1507 model, and the overwhelming reason was that it depicted all 360 degrees of longitude, and it was scientifically plotted. The plane-chart was a professional instrument of limited scope which continued to flourish in the form of the private manuscript, either for use at sea, or as an artistic commission from monarchs or nobles. Printed maps from c.1510 to c.1580 almost invariably employed an oval projection which may be thought of as the completed Ptolemy. Waldseemüller's 1516 map (and direct copies of it) is the sole example of a printed sixteenth-century world map on the plane-chart model, and it was without future influence. The 1507 map however was seminal in many ways, and its depiction of the Pacific proved prophetic. Waldseemüller's importance in creating these twin emblems from a crucial transitional period in map history can scarcely be over-emphasized. We can only regret that he died so young, that we know so little of him, and that he was not permitted to develop his geographical ideas after the age of Magellan.

(This map has been coloured in the style of a sea-chart of the period.)

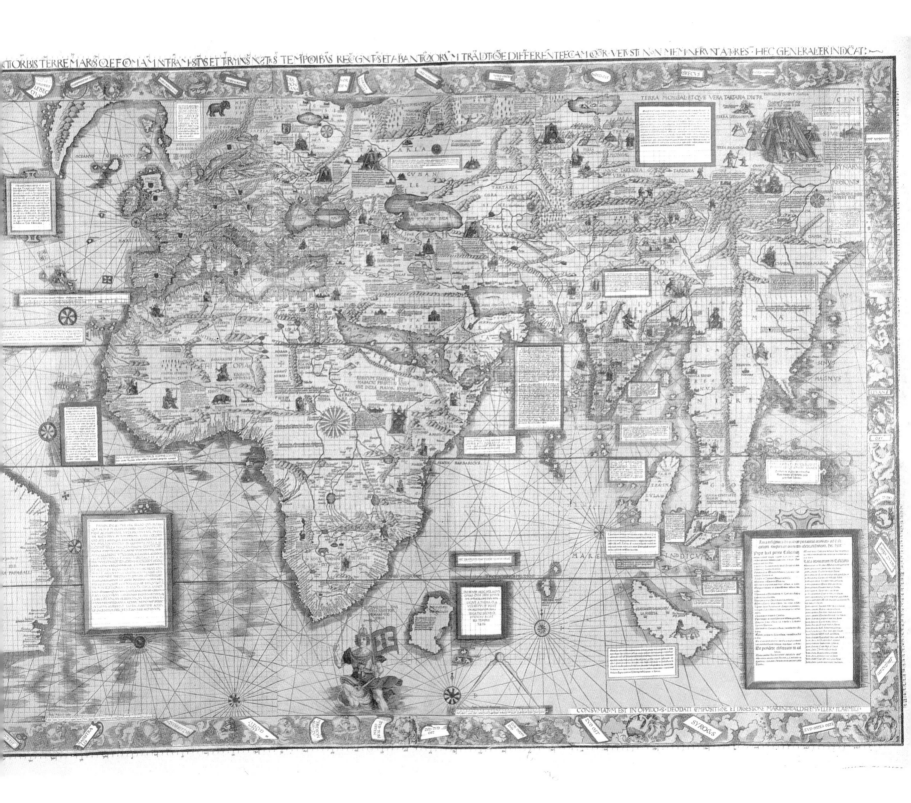

Peter Apian,
1530

THIS MAP OPERATES ON TWO distinct levels. On the one hand it is a Renaissance *jeu d'esprit,* a play of the intellect, linking the most inward central point of the individual with the most far-reaching symbol of all that lies outside the individual — a map of the whole world. But it is also a logical coherent image, not a fantasy, and there is ample evidence that it was constructed on a mathematical plan.

The texts on the map are formal and dedicatory, providing no clues to the method or purpose behind the map's construction. However the key is to be found in the outlined central portion of the map, containing Europe, western Asia, the Indian Ocean and Africa. This is in fact Ptolemy's world map, using his second projection as in the Ulm edition of 1482. This is made explicit by the small portrait of Ptolemy in the top left corner, displaying his map within an otherwise empty heart-shaped frame. The Dürer-Stabius map of 1515 was precisely the same map, but drawn in visual perspective on a globe. The Apian map however extends and completes that Ptolemaic map, retaining the same projection, but extending the parallels to 360 degrees and the meridians to the poles. The centre of the map, where the axes intersect, is precisely that of Ptolemy — in the Gulf of Oman where the 24th parallel crosses Ptolemy's central meridian (which on his map was 90 degrees east of the Fortunate Isles). It seems virtually certain that the heart shape was discovered by chance as the cartographer experimentally extended Ptolemy's grid lines. The parallels thus become a series of diminishing circles around the north pole, while the meridians become widening ellipses around the vertical axis. The apparent tilt of the globe is caused by Ptolemy's choice of 24 degrees north as the central parallel instead of the equator.

Having extended Ptolemy's grid lines, how did Apian complete the map itself? Again this is made clear by the figure of Vespucci in the top right corner: he displays a map of the new world, surrounding the blank Ptolemaic core, thus precisely counterbalancing the figure opposite. There can be no doubt that Apian had studied closely the Waldseemüller map of 1507, since the geography of this outer map section is closely modelled upon it, and Vespucci was Waldseemüller's major source. Moreover these twin portraits are also found in Waldseemüller's map, with the clear inference that Ptolemy is the geographer of the Old World, Vespucci the discoverer of the New.

Although not a visual joke in the way that the later Fool's Cap map is, Apian's heart map is a deliberate intellectual conceit. The heart's physiological function was not understood until the seventeenth century, but its shape and its association with the blood were known to ancient and medieval science. Many cultures, in Europe and beyond, had identified the heart as the seat of the emotions and of the personality. Innumerable passages from Biblical literature, from Dante, from secular romances and devotional works, all served to create the resonance of the heart metaphor. In depicting the world in the shape of a heart, was Apian consciously fashioning a symbol of man's identification with his world, suggesting a correspondence between the inner and the outer realms, between the individual and the universal? This must be speculation; ultimately the metaphor is elusive. But the force of intellectual play has been here directed at shaping a new image of the world, a process wholly characteristic of the Renaissance, and one which was to engender a richness unequalled in the history of mapmaking.

(This map has been coloured here in the style of the period.)

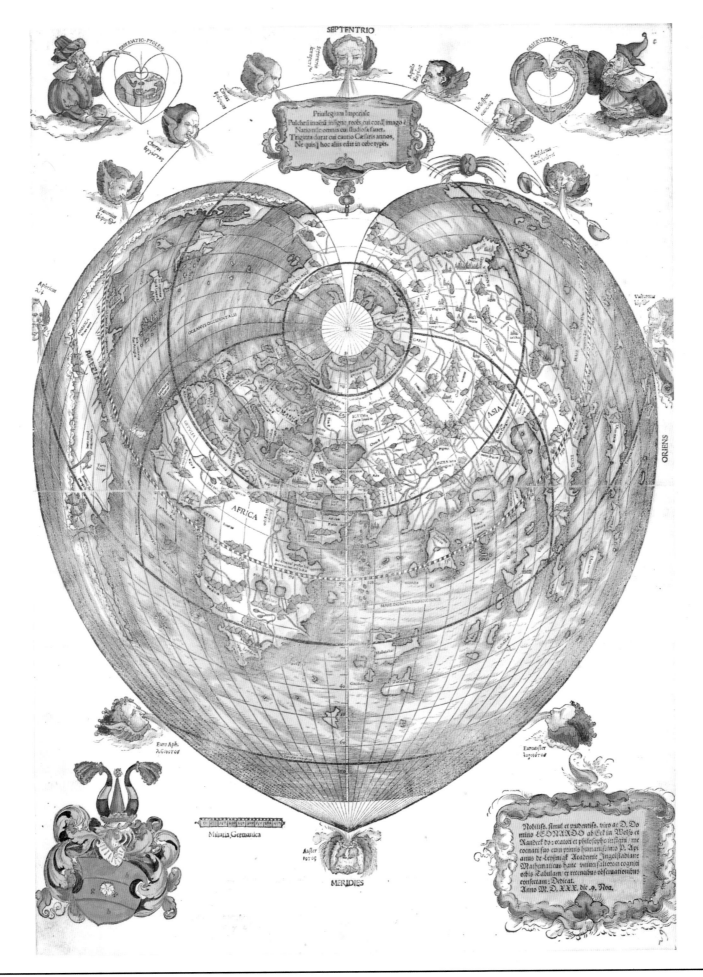

Battista Agnese,
c.1536

In the history of mapmaking the manuscripts of Battista Agnese are probably the nearest direct approach to artistic work, in both their form and their context in Italian society. A native of Genoa, active in Venice between c.1535 and 1565, he is known to have produced one hundred or more manuscript atlases, usually containing eight to ten maps each. In their miniature format and elegant execution they are comparable to contemporary manuscript illumination, which survived and flourished for almost a century after the advent of printing. Agnese's atlases were bought by wealthy clients and the number that have survived — more than sixty — indicates the value, artistic and financial, that was attached to them.

The majority of the maps in the atlases are sea-charts in style if not in reality. They show no land features, only sequences of coastal names, and a network of navigation lines. They are unadorned by visual imagery or literary texts; they are empirical documents, showing features known to, and verifiable by, seafarers. Agnese's sources for these maps are not documented, but he must have had sources of geographical information in the maritime centres of Europe, principally Portugal and Spain. His maps were the first to show the peninsula and gulf of California, and they were not static over the three decades of his productive career. However his maps are rarely dated, and no attempt has yet been made to plot the precise interrelationship of the atlases, with a view to measuring their value as a record of European discovery and expansion.

These sea-charts are drawn on the plane-chart model, that is a notional rectangular grid covers the map, whose latitudinal and longitudinal sides have a constant ratio over the whole map. Beyond 70 degrees north and 60 degrees south, the map dissolves into emptiness, as it does also in the Pacific. Since the map is not a mathematical projection, it has no framework; one understands that it is not strictly a map of the whole world, and its empty edges are visually acceptable on that basis. However, the significance of the Agnese atlases for our understanding of the evolving world map lies in the addition of a second, very different world map. These maps are on an oval projection and they show 360 degrees of longitude. They show land detail, and coasts beyond the limits of maritime exploration are completed. Their role as a map of the entire world is underlined by the tracks of Magellan's circumnavigation of 1519–22; indeed this became a hallmark of the Agnese maps.

How should we understand the appearance side by side in the same atlas of two distinctly different, even conflicting, maps of the world? The answer is that the sea-chart clearly represented reality as far as was known. But the intellect also demanded a conceptual image of the entire world to complement the partial truth of the sea-chart. The conceptual map contains many Ptolemaic elements even where these conflict with contemporary knowledge, such as the entire delineation of southern Asia. It is therefore a world map from within the scholarly tradition of the late medieval and Renaissance cosmographer, whose role was to distil an image of the world from many sources — classical literature, rational speculation, travel narratives, geometric science. Such a tradition could not rest content with empirical coastal outlines. The intellect sought a lucid, concentrated image of the whole world, and this the cosmographer not the navigator could provide. This duality of scholarship and practicality, theory and empiricism, would have been understood by contemporaries without provoking an insistence that one was right and the other wrong. The play of intellect could hold both in balance.

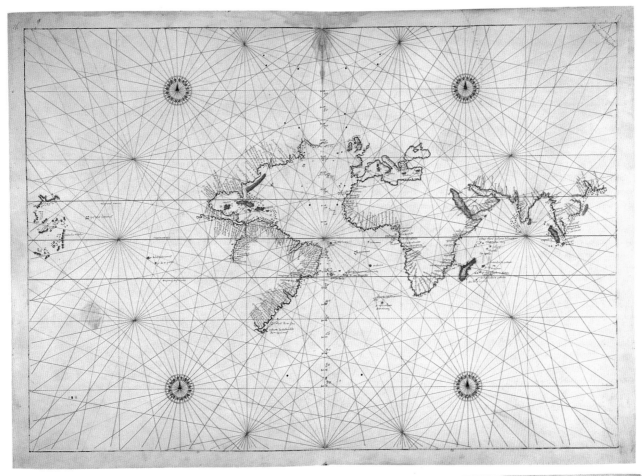

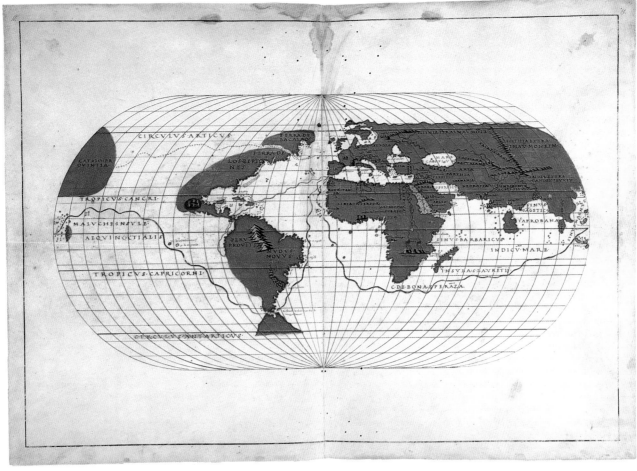

Jean Rotz, 1542

THIS MAP IS THE EARLIEST known example of the double-hemisphere construction, which, in various forms, was the dominant model of the world map throughout the whole of the seventeenth and eighteenth centuries. Rotz's version is a sea-chart in character: it shows coastal outlines only, and the relative emptiness of the eastern hemisphere obscures the virtues of the twin-hemisphere structure, which later became clearer and more convincing as the map became fuller. Rotz was a mariner and chartmaker of the school of Dieppe, but he spent some years in England where he put the finishing touches on his highly-wrought manuscript atlas of sea-charts, *The Boke of Idrography,* dedicated to King Henry VIII, which contained eleven regional charts and this world map. There is no textual commentary on the maps, and we can only conjecture what motives drew Rotz away from the conventional plane-chart to this highly original projection.

Each hemisphere is graded for 180 degrees of latitude and longitude, thus the whole world map is represented. Its obvious advantage over the single-sphere oval world map is its reduced distortion, as the meridians curve less sharply. At first sight it is surprising that such a projection should appear in a sea-chart atlas since it is in reality not one map but two, and navigation across the interrupted areas is impossible since they do not exist geometrically. Each hemisphere is projected onto a plane which is conceived to touch the globe at a focal point on the equator. The lines of projection radiating from that point to the tangent plane are clearly marked. These lines have the property of showing true direction from that point, and the linear scale is accurate along those lines. Thus each hemisphere becomes effectively the visual equivalent of a huge compass, and has valuable navigational properties. This must have been the intention in Rotz's mind when he created this innovative projection, despite the limitation arising from its interrupted structure.

The map's geography is largely restricted to known coastlines. It is perhaps surprising to recall that of all the regions of the world, northern Europe was among the least known: this map antedates by eleven years the first voyage around the North Cape to Archangel. Scandinavia here ghosts into empty sea, as do North America and East Asia. This empirical restraint is carried to such length that a ship appears to sail north of India. By contrast a swollen land-mass identified as the Land of Java was purely hypothetical, a concept peculiar to the Dieppe school. It was presumably based on unconfirmed or confused reports of Portuguese landfalls, and its significance vis-à-vis Australia has been much debated by historians of exploration.

Rotz's map was a manuscript which remained in private hands, unseen by the major mapmakers of Europe: it was without influence and it cannot be regarded as the model for the later generation of twin-hemisphere maps. But what one man can discover, another can re-discover. The visual symmetry and intellectual appeal of a world map split into an eastern and a western hemisphere, the old world and the new, were felicitous discoveries which the play of intellect would rediscover in subsequent decades, while Rotz's map lay unstudied in the royal library. Half a century on, when the coasts of America and Asia were well defined, the twin-hemisphere map elegantly fulfilled its promise, prefigured here by Rotz.

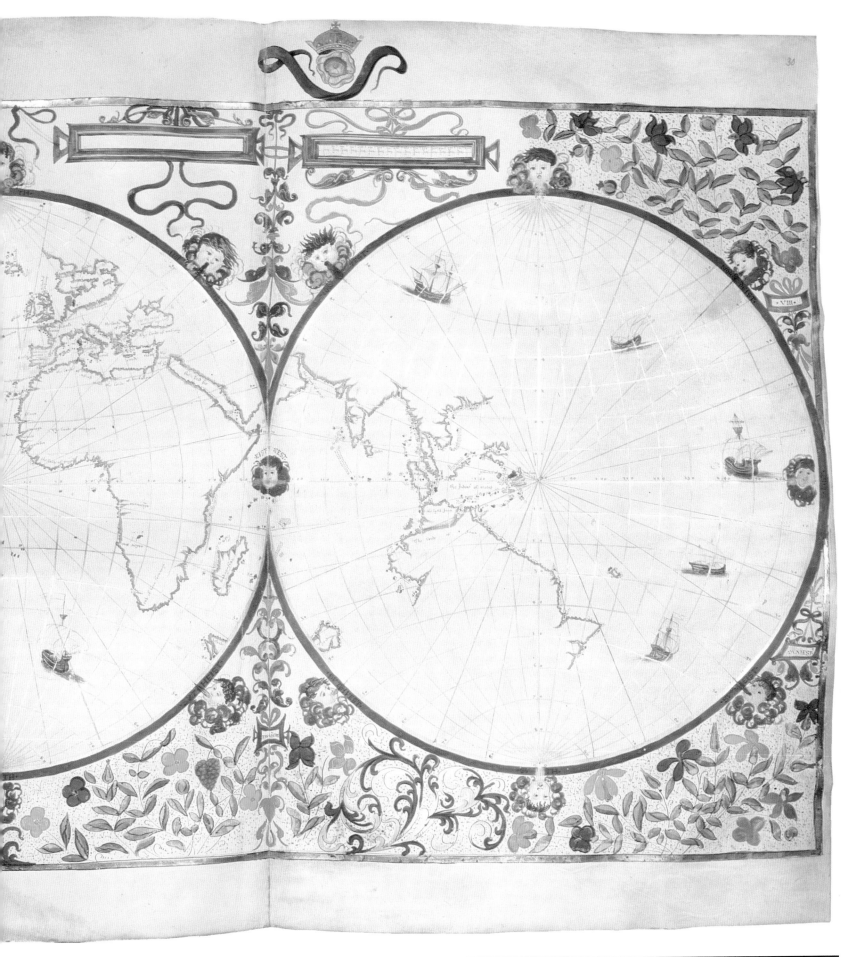

Pierre Desceliers,
1550

For more than a century after the application of printing to maps, c.1475, manuscript world maps continued to evolve in a different context. In the early years of the age of discovery, Spanish and Portuguese manuscript maps embodied new knowledge which was treated as a species of state secret. But despite national and commercial rivalries, geographical secrecy was possible for only a limited time. Nothing could prevent other explorers and other nations from exploiting the freedom of the seas. By the mid-sixteenth century French ships were sailing regularly to the Newfoundland fishing grounds, and Canada was being seen as the French *Terre Neuve*. A school of chart-makers became active in Dieppe, producing some of the most detailed and ornate charts in Europe, some for use at sea, others richly embellished commissions for wealthy patrons. Pierre Desceliers was one of the leading figures of this 'Dieppe school'.

The Desceliers map is firmly within the nautical plane-chart tradition. Its most striking visual aspect is that it has two orientations: all texts and pictures north of the equator are inverted. From this we must infer that it was intended not to be hung on a wall, but to be spread out and studied on a large table, so as to be right-reading from both sides. The map is a 'Plane-Chart', that is one constructed not on a true projection, but on a simple square grid, the ratio of the longitude to the latitude degree being equal at 1:1 across the whole map. This gives the map its rectangular structure, generally resembling what we now know as the Mercator projection. However on this principle no allowance is made for the lessening value of the longitude degree. Equally distorting is the omission of 90 degrees of longitude, covering the Pacific Ocean. After the great Magellan voyage of 1519–1522, more than four decades were to pass before the Pacific was crossed again. It was known to be of vast extent, but no one knew the coasts which bounded it, nor what islands it might contain. A plane-chart of this date would be dominated by the great vacuum of the Pacific, therefore Desceliers omits it.

These structural and geographical limitations make it clear that the Desceliers map, despite its compass-roses, wind-faces and navigation lines was a sea-chart in style only. It was no doubt more important to its creator and to its patron as a conceptual image of the sixteenth-century world, expressed through text and image. There are more than fifty vividly-painted tableaux and twenty-five extensive texts on the map, and they are a curious mixture of modern history and medieval myth. Early French attempts to colonize Canada are described, the conquest of Peru by the conquistadors, and the Portuguese East Indian trade. But alongside these we find descriptions of legendary Amazons (in Russia), pygmies at war with cranes (in Canada) and the legendary king Prester John (in Ethiopia). As the progress of exploration failed to discover any reality corresponding to these legends, they were re-located to other, still unexplored regions. Their inclusion on a map which summarizes the achievements of the Age of Discovery demonstrates the force of tradition in mapmaking: even while presenting a picture of the contemporary world, the mapmaker wished to affirm his place in the cartographic tradition by retaining the intellectual landmarks of the past. In many ways this is clearly a Renaissance version of the medieval *mappa mundi*, a visual encyclopedia of kings and peoples, flora and fauna, folklore and scholarship. Within two decades of the drawing of this map, the cartographic world was to undergo radical changes in the hands of the Flemish mapmakers Ortelius and Mercator. Their technical and commercial innovations placed mapmaking on a different plane from that of the private manuscript tradition, and maps like the Desceliers, though not yet extinct, would become magnificent anachronisms.

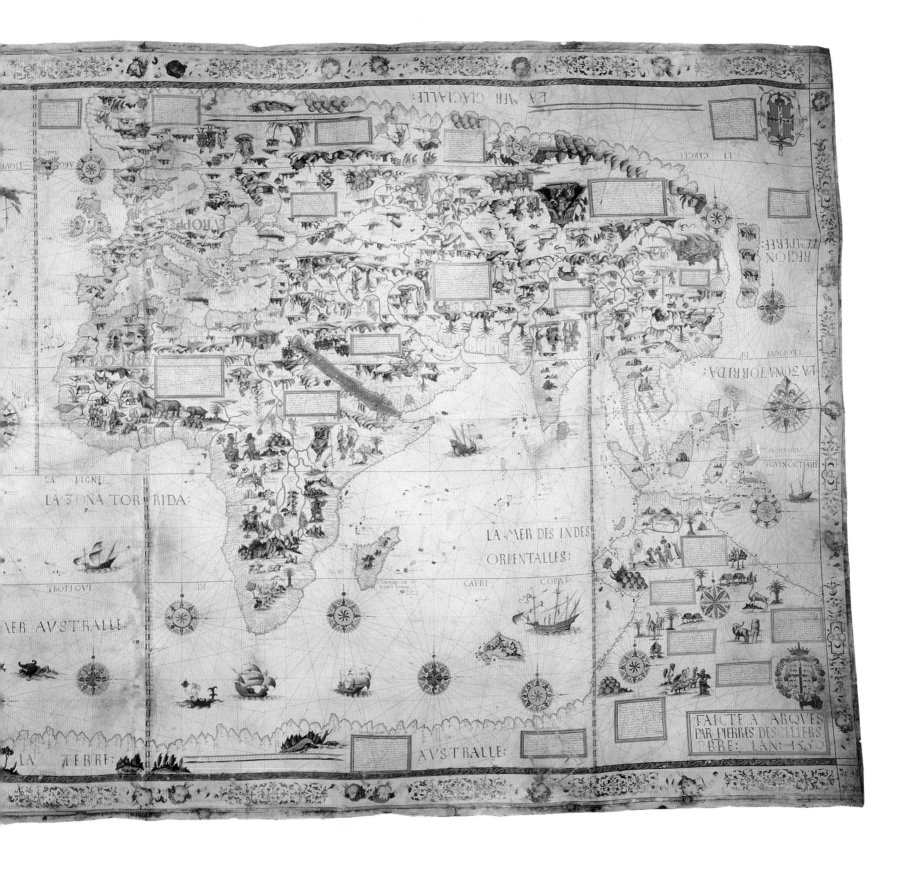

Giovanni Camocio, 1567

ITALY HAD BEEN AT THE forefront of intellectual and technical innovation in mapmaking, from the first editions of Ptolemy in the 1470s through to the seminal work of Rosselli. Map printing in Italy subsequently declined in volume and innovative quality, while in the hands of a small group of artists like Agnese, the manuscript map enjoyed a revival. Venice emerged as the centre of Italian mapmaking in the 1540s, with a school of distinctive and prolific cartographers. A succession of world maps, woodcut and engraved, was issued by the leading figures, Gastaldi, Forlani and Camocio, which satisfied a wide market demand throughout Europe before the rise of Flemish map publishing.

Giovanni Camocio's world map of 1567 is entirely characteristic of this school. The oval projection, derived from Rosselli's archetype at the beginning of the century, was now dominant throughout Italy and Germany. Camocio's map differs slightly from Rosselli's in that the poles are artificially represented by an extended line, enabling the meridians to terminate without excessive curvature. The four corners created by the oval projection naturally lent themselves to decoration with artistic motifs, in this case figures personifying the four elements. The geographical hallmark of the Venice school is immediately apparent: the gigantic southern continent, reaching here to within 20 degrees of the equator, named disingenuously *Terra Incognita,* or *Discoperto de Novo* ('recently discovered') or, reaching back 300 years to Marco Polo, *Terra Lucach.* The appearance of this hypothetical land-mass on sixteenth-century world maps and its possible source in Portuguese voyages and literature, have long been studied as conceivable early evidence of encounters with Australia. However the contemporary accounts (of figures such as Barbosa, Pinto, or Serrao) are just not precise enough to draw any firm conclusions. The ultimate sources for this *terra incognita* are almost certainly literary, and not empirical at all. The Ptolemaic concept of a southern continent to counterbalance those of the north, and the endlessly-influential reports of Marco Polo concerning the lands south of China (variously named Beach, Lucach or Maletur) combined to produce an expectation, a desire, for the discovery of yet more new continents.

Perhaps the most significant single innovation on the Venetian maps of this period is the appearance of the 'Strait of Anian' (a name taken from Marco Polo) separating Asia and America. Magellan's voyage of 1519–22 across the southern waters of the Pacific, did not in itself settle the question of the relationship between Asia and America. The possibility of a northern link remained, and many sixteenth-century maps affirmed its existence. However after the conquest of Mexico in 1521, the Spaniards explored steadily northwards into the southern regions of North America. De Vaca, de Coronado and de Soto travelled by land, de Ulloa, Alarcon and Cabrillo by sea, through the modern California, Arizona, New Mexico and Texas. These lands and peoples were unrelated to any strands of European thought or tradition about Asia, even to the writings of Marco Polo himself: there was no evidence, physical or cultural, that this was indeed the Asia that the first explorers had sought and had taken it to be. The conclusion became irresistible that this was a new world, and the fact was registered by the appearance of the hypothetical Strait of Anian. This was an important psychological moment in the perception of the world as now divided into five continents in two distinct hemispheres. It made possible the splitting of the oval world map into the symmetrical and fitting twin-hemisphere image. The Venetian school of mapmaking did not long survive competition from the emergent Flemish cartographers, but it bequeathed several important legacies to Ortelius and Mercator, the most far-reaching being the Strait of Anian, the final separation of the Old World from the New.

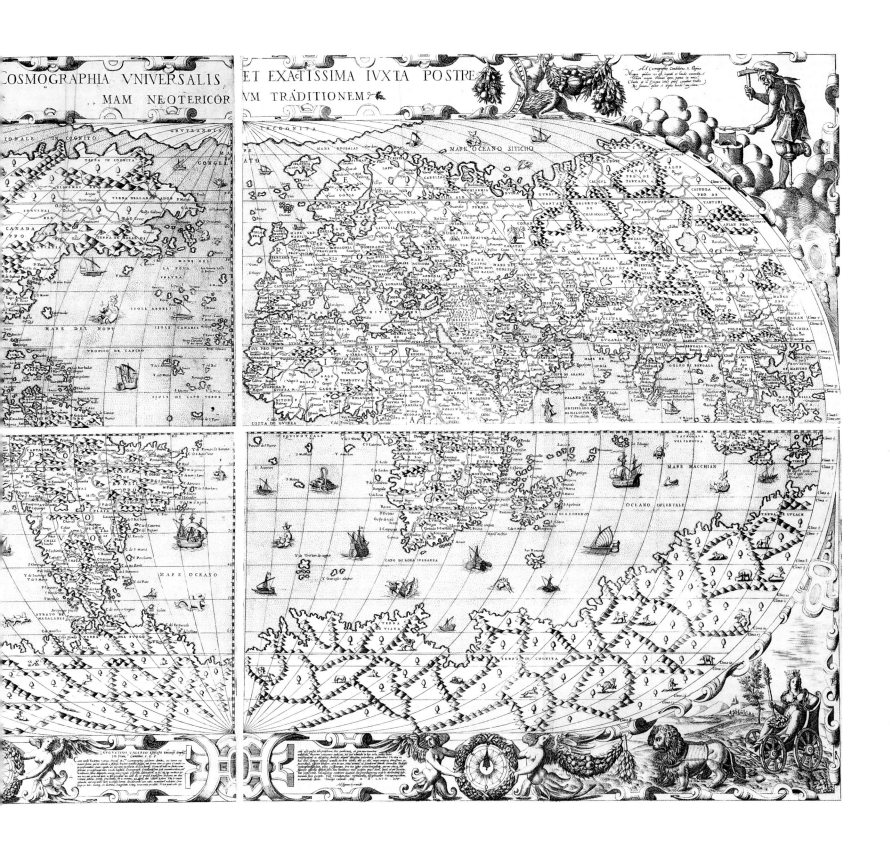

Gerard Mercator, 1569

I N THE HISTORY OF MAPMAKING Gerard Mercator (1512–1594) is a towering figure, his stature comparable with that of Ptolemy. He brought a new scientific rigour to the editing and production of maps, and as a publisher he made commercial innovations which opened new markets for all who succeeded him. A complex and gifted man, trained for the priesthood, he conceived a passion for geography, and by his twenty-fifth year he had mastered the contemporary canons of mathematics, geography and astronomy, and was a highly-skilled engraver, calligrapher and instrument-maker. A deeply religious and cultured man, intimate with classical and Biblical literature, he saw himself as a Renaissance cosmographer, one whose task was to present a coherent, rational interpretation of the world. Settled in Catholic Louvain, he was imprisoned for heresy by the Inquisition, and after his release he transferred to Duisburg in the Protestant Duchy of Cleves, where he worked for almost half a century on his grand conception of the world atlas, which remained only partly complete at his death. He also wrote Biblical commentaries and a treatise on lettering, in which he developed the cursive italic script which helped to distinguish his maps. During the latter half of the sixteenth century, the centre of map publishing shifted north from Italy to the Netherlands; in that process Mercator's personality was a key factor, and his achievements formed one of the foundations of Dutch mapmaking supremacy in the seventeenth century.

After producing an important globe and publishing his first world map in 1538, Mercator published no other world map for thirty years. During this long period he researched deeply in geographical literature, travel narratives and the techniques of navigation, all in preparation for this great world map, which was revolutionary in scope and structure. That Mercator conceived his map as an intellectual document to be read and studied, not merely to be looked at, is clear from the many lengthy texts which the map contains. These texts constitute brief essays on disparate subjects: the disposition of the earth's land-masses according to classical and contemporary theory; the technique of direction-finding and mensuration; the history of the peoples of Asia, with a demythologizing account of the Prester John legend; an ornate panegyric on the felicity of his adopted home, the Duchy of Cleves. Taken together, these texts are a revealing index to late Renaissance geographical thought: despite the excitement of the new discoveries, its overwhelming concern was continuity, a dialogue with the authorities of the past.

The most significant text, in the eyes of history, is that (in North America) which explains the novel basis on which the map has been constructed. For over two centuries the plane-chart had been accepted as the sea-chart, the navigator's chart, the empirical chart, and it had retained its place through all the changes to the conceptual world map which took place between 1450 and 1550. Its virtue was that, in the Mediterranean world where it originated, distances could be measured fairly accurately, and in that restricted region its defects passed unnoticed.

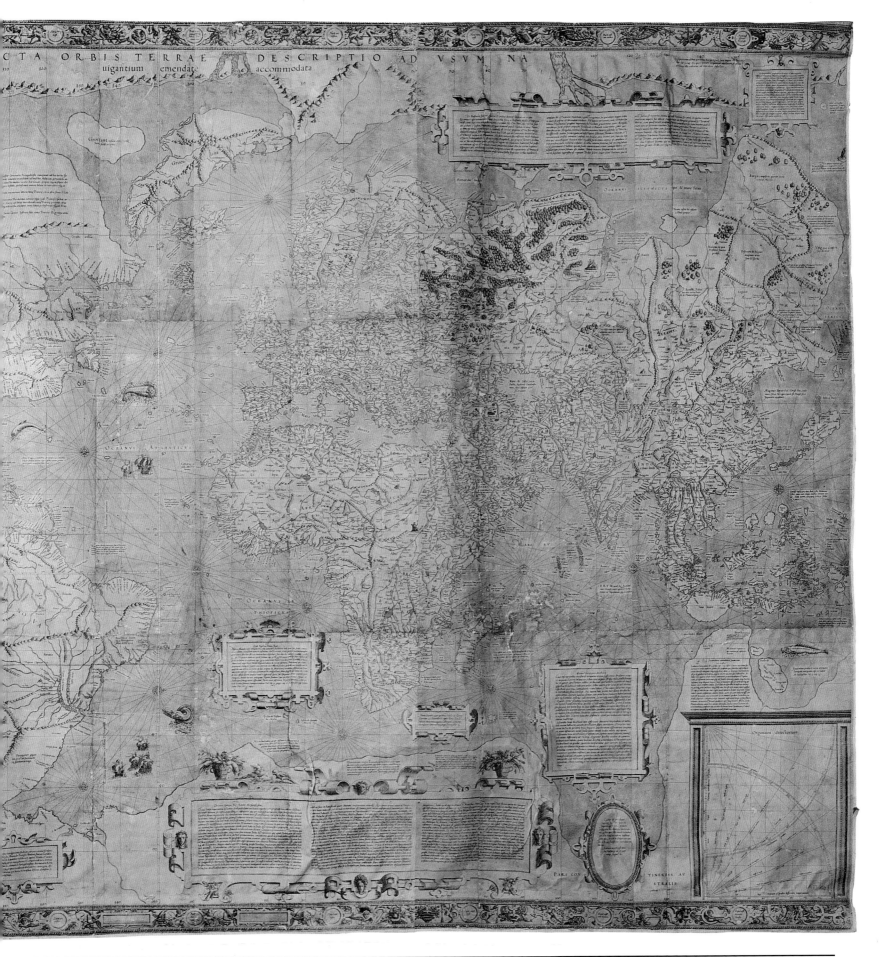

But as the world expanded, and the mariner crossed oceans and hemispheres, it became an anachronism. Mercator was the first cartographer to rationalize its problems, and the first to solve them. He gives a succinct diagnosis of the fatal flaw of the plane-chart: in fixing the ratio of latitude to longitude as a constant across the whole map, the plane-chart makes no allowance for the diminishing value of longitude towards the poles; in effect it ignores the fact that the earth is a sphere. A map of the earth cannot simply be spread on a flat surface with degrees of latitude and longitude plotted as a pattern of regular squares. In Mercator's map the lines of latitude and longitude are straight lines intersecting at right angles, just as on the plane-chart. The difference is this however: meridians on the globe converge at the poles, hence as the plane-chart shows them as parallel lines, they are really widening. To balance this, Mercator reasoned that it was necessary also to increase progressively the value of the *latitude* degree towards the poles. Thus the ratio between latitude and longitude degree remains accurate throughout the map and the linear scale is increased. The equatorial scale gradually increases northwards or southwards until at the poles, could they be shown, it would be 1:1 and the map would be reality. This gives the Mercator world map its distinctive appearance of being stretched towards the north. This is less obvious on this original prototype than on the modern map because the Arctic regions were still unexplored.

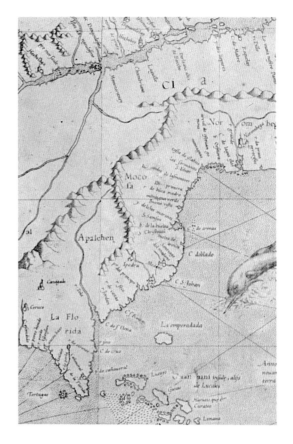

What is the practical result of this calculated geometrical construction? It is often stated that the Mercator projection 'shows true direction', but this is inaccurate. A straight line on Mercator's map is not the shortest distance between two points on the globe, as it would have to be if that statement were true. What is true is that a straight line on Mercator's map is a line of constant bearing. The mariner navigating from place to place could read a single bearing along that one line, set his compass, and arrive at his destination, after making necessary allowances for winds and tides. In reality, on the globe, such lines of constant bearing (rhumbs) intersect each meridian at a different angle (except for due east and due west) and the compass bearing to north was constantly changing. Therefore on the non-Mercator chart, direction-finding necessitated regular changes of bearing as each meridian was crossed, in order to trace a true rhumb-line. The genius of the Mercator map is to make such rhumbs appear as straight lines: what you saw on the map was what happened in reality. For the purpose of navigation this is obviously highly desirable. It is true that the projection also has a number of defects arising from its variable scale. The first and most obvious visually, the distortion in the relative sizes of the land-masses, is of little practical concern to the navigator. Moreover this defect is lessened when the Mercator projection is applied to its proper object in a series of regional ocean charts. On the regional scale the benefits of Mercator's system emerge more clearly; the world map should probably be regarded as a paradigm of the projection. More serious, especially in the eyes of the mariner, is the fact that distances cannot be readily measured, unless the course happens to be directly along one parallel, because the scale is not constant. However in the lengthy text placed west of South America, Mercator ingeniously explains how distances can be measured by converting geometrically the length of the bearing line into equatorial degrees, whose value is known and fixed.

That Mercator's map was ahead of its time is indicated by its lack of immediate influence. Mercator did not choose to give detailed instructions for plotting the grid system, nor he did publish a large-scale chart to demonstrate its direction-finding qualities. There is no doubt that Mercator's sophisticated approach to geometric projection was understood by few of his contemporaries. For more than a century after this map's appearance, sea-charts and maritime atlases continued to be published which ignored the problem of projection. A handful of Mercator-projection maps were published during the seventeenth century, but it was almost totally eclipsed as a conceptual image of the world by the twin-hemisphere model. Even Mercator's son Rumold re-published his father's map re-plotted to the double-hemisphere configuration in 1587.

The structural aspect of the Mercator map has overshadowed its other features, and geographically it is an eclectic and revealing document. The arctic continent shown here is pure medievalism, the legend of the land of four rivers narrated by a fourteenth-century

English monk, Nicholas of Lynn, who claimed to have travelled there; this fable does not appear on any surviving medieval maps, but for some reason it was taken up by sixteenth-century cartographers. The Arctic Oceans so clearly visible on this map were influential in stimulating the attempts to find the northeast and northwest passages to the orient. The antarctic continent is pure hypothesis, again based on the traditional authority of Ptolemy and Marco Polo. On the other hand there is a distinctly contemporary enigma on this map: Mercator is the first to mark a large inland lake in North America, in contrast to his contemporaries Ortelius or the Venetian school. The Great Lakes were not 'officially' discovered for a further forty years. It appears here too far north in relation to the St. Lawrence, and the mystery is deepened by Mercator's note *Mare est dulcium* — 'This lake is of fresh water'. Again, in Africa Mercator departs from Ptolemy in displacing the sources of the Nile far to the west and south, but on what authority we can only guess. The huge distension of North America can probably be explained largely by the notorious difficulty of determining longitude. The separation of America from Asia is decisive. The editorial process that went to create a world map in this era, even for a serious scientist like Mercator, still involved wrestling with totally unverifiable sources and authorities.

The ultimate significance of Mercator's map lies in its scientific creativity. After the discrediting of the medieval world image, charts such as the Genoese and the Cantino display a sharp awareness of empirical reality. Ptolemy's direct influence in the field of projections was limited, leading to such experimental images as the heart-shaped world. Waldseemüller and Rosselli made the crucial advances in displaying the entire world, but the oval projection remained a visual solution, elegant but navigationally useless. At the same time the professional conservatism of the mariner bound him to a traditional form of sea-chart which was deeply flawed. Mercator was the first cartographer consciously to re-shape the world map as a mathematical exercise with a desired end. He was not concerned with visual appeal, and what he created did not claim finality: it fulfilled a precise technical purpose, and as such it was a triumph of scientific virtuosity. It was destined to transform the printed image of the world, but not yet.

Antonio Saliba, 1582

'Come Mephistophilis let us dispute again,
And reason of divine astrology.
Speak, are there many spheres above the moon?
Are all celestial bodies but one globe,
As is the substance of this centric earth?'
— Marlowe: *Doctor Faustus*

THE MAP OF THE EARTH'S surface was the most urgent and practical problem for the geographers of the sixteenth century. But there was another dimension to the puzzle: what was the space within which the earth existed, and what lay beyond it? What were the clouds, the winds, rainbows, thunderstorms? What were the stars, the sun and the moon, and how did they move? Heaven and hell were known to be above and beneath this world, but where exactly, and how did they co-exist with the observed realities of sky and earth? The complex of cultural changes which we call the Renaissance was artistic, literary, political and many other things, but it was not scientific. The principles and methods of empiricism — observation, experiment, deduction — lay still in the future. Even the Copernican revolution, profound though its implications were, was not based on methodological innovations, and it provided answers to none of the above questions. To visualize the earth in space, the only model available was still a system of concentric rings, and, since Copernicus' theory was not widely accepted until the century after his death, the centre of the system was still the earth.

This cosmological chart, which appears so bizarre and unfamiliar, is in fact only mildly unorthodox as a pre-scientific image of the cosmos. The author was Antonio Saliba, a Maltese about whom nothing is known save his own claim to be a doctor of theology, of philosophy and of canon law. The work's title promises to display 'All things which are in the world and in the heavens, for the universal benefit of all who would know the occult secrets of nature'. The world map, restricted to about 50 degrees of latitude, appears in the middle circle, having four inner and four outer rings. The circle immediately within the earth shows the process of Judgement, the separation of the saved from the damned, next the circle of purgatory, then the outer and inner circles of hell. Beyond the earth are two circles of aerial phenomena — storms, rainbows and the twelve classical wind-faces. It is in the eighth circle that Saliba's unorthodoxy and occultism are given freest rein. The appearance in 1577 of a great

comet (now named Tycho Brahe, and considered to be the brightest ever recorded) was, inevitably, interpreted as having prophetic significance. Saliba seems to have regarded this so seriously that the eighth circle is devoted largely to descriptions of comets, their historic appearances and occult significance.

The cosmic model of concentric rings was derived from Aristotle and Ptolemy, and, in various evolved forms, it prevailed until the seventeenth century. The Ptolemaic model comprised nine spheres around the earth: five planets, the sun, the moon, the stars, and the *primum mobile*. The spheres were conceived to be made of crystal, into which the heavenly bodies were fixed, with the spheres revolving upon a common axis. Christian teachings required inner realms in which to locate purgatory and hell, while additional spheres of fire and air were sometimes added. Outside the whole system was the empyrean, the tenth heaven, the abode of God, the angels and the saints, though this was not to be conceived as fixed within spatial limits. This is the cosmic structure underlying Dante's *Divine Comedy*, of nearly three centuries earlier. Saliba's departure from the classical content of the nine spheres while retaining the structure, is entirely typical of the fluid state of Renaissance science. The spheres of the sun, moon, stars and planets must be conceived as compressed into the eighth sphere, dominated by the comet; this is Saliba's most unconventional step. The depiction of the ninth heaven as a circle of empyrean fire reflects the Renaissance hermetic reverence for fire as the purifying principle, through which nature could be transformed and made to yield her secrets.

Saliba's visualization of the earth is a kind of cross-section of the physical, spiritual and conceptual world of the sixteenth century. How literally was such an image regarded by his contemporaries? The world map is real enough, even if it is patrolled by demons, and in the absence of Newtonian physics, no other model was available. The armillary sphere, the most familiar contemporary scientific instrument, was a linear version of the same structure. Saliba was attempting here a subject at the limits of possible visualization: the difficulty of projecting an image of the cosmos clearly exceeded even those of projecting the world map. But the process of modifying a classical model is strikingly similar, as is the task of deciding which traditional doctrines to retain in the light of contemporary knowledge. Saliba may not have kept his promise to reveal the occult secrets of nature, but he has added another dimension to the world map.

ANTONIO SALIBA, 1582

CHAPTER FOUR
The Theatre of the World

Early in the seventeenth century, Francis Bacon identified three great technical innovations as the distinctive achievements that had shaped his age: the printing press, the magnetic compass and gunpowder. This clear, almost brutal insight into the technological basis of European power would be amply justified during the next three centuries. In the century in which Bacon wrote, the political and commercial prizes of the Age of Discovery would be contested in extra-European adventure, war and trade, and that contest shaped a new world image.

Magellan's first circumnavigation of the earth in 1519–1522 completed in some senses the Renaissance perception of the globe. The Pacific Ocean, its unsuspected vastness revealed by Magellan, had been the missing section of a mysterious jigsaw. But much remained obscure. America was clearly a new world and not Asia, but a northern link between the two might still exist. Was there also a northerly route from the Atlantic into the Pacific? Was there another undiscovered world to the south, of which Tierra del Fuego and Java were the northern coasts? Magellan's feat was so far ahead of its time that it was not attempted again for more than half a century. When it was repeated by Drake in 1577–80, and by Cavendish in 1586–88, the result was a renewed sense of 'The World Encompassed'. The spherical globe was no longer merely a concept, but had been placed in the realm of human experience, so that the entire earth was perceived as the world of man, specifically of European man, open to be explored and possessed.

The entry of the northern European nations into the age of exploration was slightly delayed, but when it did occur it undoubtedly contributed to the late flowering of their Renaissance. This was a decisive moment in European history, when intellectual curiosity, technological strength, national rivalries, and overwhelming avarice, all combined in the European mind to transform the world into a theatre, where aspirations of wealth and power might be realized. At this time too the centre of gravity in European mapmaking shifted northwards from Italy to the Netherlands, and the twin-hemisphere map appeared and flourished. The single-oval projection had dominated the sixteenth century, but with the conjectural placing in the 1560's of the 'Strait of Anian' to separate Asia from North America, the moment was clearly ripe for the single sphere to be split into two, producing the symmetrical, balanced image of the old world and the new.

The transformation in the power and fortunes of the Netherlands proved to be crucial to the history of mapmaking. In addition to being positioned at the trading crossroads of Europe, the Dutch through their sea-power succeeded in wresting from the Portuguese vital sea-routes and trading bases in Africa, South Asia and the East Indies. Amsterdam became the centre of international trade and finance, and there followed the cultural flowering known as the Dutch Golden Age. The presence in Amsterdam of scientists, seafarers, artists and craftsmen, lies behind the sudden maturing of commercial map publishing, where a dozen rival publishers sought to dominate this rich market by producing ever larger, more detailed and more artistic maps. The ornamented double-hemisphere world map became the archetypal image of that culture, as the mapmakers exploited its expressive possibilites to create a superbly burnished mirror of their own society. A storehouse of graphic motifs was built up which could be manipulated at will, motifs drawn predominantly from the natural world, but mediated through the forms of classical literature and art. The four elements, the four seasons, the gods of sky and sea — these were the favourite images. Portraits of historical monarchs, scientists or navigators, such as Caesar, Magellan or Ptolemy, were also used, and contemporary royal portraits. Less frequently the seven wonders of the ancient world appeared, or contrasted

The deities Mars, Jupiter and Apollo still decorating the rationalized world map of the late seventeenth century. From de Wit's world map, page 95.

Romein de Hooghe: a political allegory c.1705. The Spanish Pretender Charles III is offered the globe by secular and spiritual powers.

tableaux of communities at peace and war, the age of gold and the age of iron. Strikingly absent are any Christian motifs: only a handful of maps from this period feature religious or Biblical images. An approach to real geography appears in views of the world's principal cities and in the costume-portraits of national figures, the European dress contrasted with the nakedness of savages and the opulence of the east. The oceans are crowded with whales, tritons and sailing-ships, the latter always European and always fighting. These motifs appear at first sight to be drawn from very different realms and to operate on different levels — a portrait of a contemporary prince for example, beside that of Neptune impaling a sea-serpent. What unites them is the intention of the mapmaker to display not merely the world but the forces which shape and control the world. These forces are drawn within two very distinct frames of reference: first the classical, poetic tradition in which nature is personified by gods, demigods and presiding genii, and second the real-world political celebration of the power of princes and heroes to dominate the earth. There is a sense, especially in the larger maps, of baroque theatre, in which gods or monarchs survey or manipulate the human drama, as trading ships sail on beneath Apollo's chariot, or warring navies clash beneath the gaze of royal portraits. The metaphor of the theatre of life is a commonplace of seventeenth-century literature, and it is easy to see how the European mind could conceive itself to have become the principal actor in the drama of international discovery, war and trade. This theme is made explicit in maps other than the twin-hemisphere type, where the globe itself is subsumed into a political image: it may be presented to a monarch by the symbolic figures of Victory or Justice, or integrated within a royal emblem.

In order to grasp the contemporary significance of the baroque world map, it is necessary merely to analyse the elements from which it was composed: in actual or symbolic form we see cities, peoples, monarchs, explorers, fauna, peace, war, and commerce. Some of these elements are contemporary, while others are there to proclaim a continuity with the past, both real and mythical. The measurement, exploration and conquest of the earth is presented as a historical process in which the contemporary rulers inherit the role of the great protagonists of the past. These visual elements are built up into a dramatized picture of the world in its external, geographic aspects. Parallel to the spare, conceptual language of the map itself, is a symbolic language of visual imagery, displaying the world in a way that cartographic draftsmanship alone cannot. The idiom of that visual language is drawn from the aspirations of the society that created it: it is a secular language, a language consciously rooted in a largely mythical past, and a language of externals. It has no spiritual or religious dimension and the inner world is not represented. This graphic idiom revived the world map's role as a visual encyclopedia, a concept familiar from the middle ages. Vivid but unsystematic, it became a vehicle for contemporary art and aspirations, appropriate to an age of maritime adventure.

In the early seventeenth century, there was a keen sense of the heroic in these maps, and they have a dramatic, innovative excitement. But the pressures of commercial publishing debased this form of map. The motifs were mercilessly copied and re-copied, until they became stripped of any contemporary resonance. At their most expressive, these maps created a sense of the world possessed, politically through exploration and conquest, intellectually through geographical imagery. Like the medieval *mappae mundi,* their strength and their significance for contemporaries lay in their offering an interpreted image of the world, one which harmonized real geography with cultural ambitions, an image in which man felt at home. The Renaissance had produced Ptolemaic maps that were cold and geometric, and maps of discovery that were disorientating in their novelty. Is it fanciful to suggest that the great illustrated maps of the seventeenth century sought to restore the sense of man's belonging in an interpreted world, the maps now taking a secularized form, in which European man was the undisputed master of his world?

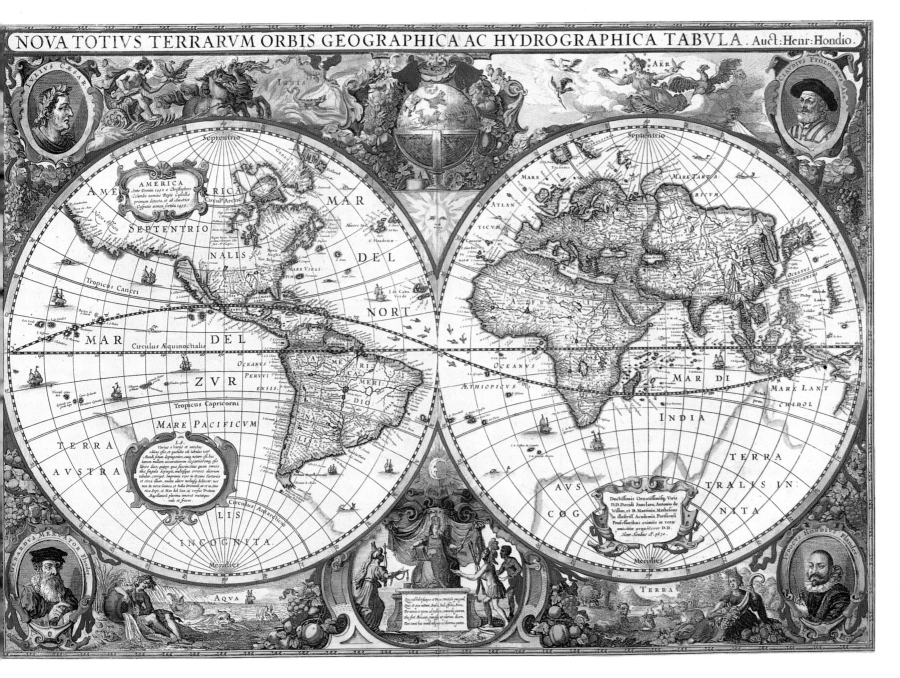

Hondius' World Map, 1630. The quintessential world map of the period: the four elements, the four continents, the sun, moon and stars surround the map, while the mapmaker's own portrait balances those of Ptolemy, Mercator and Caesar.

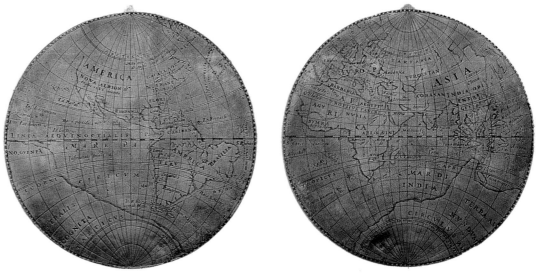

Silver medallion commemorating Drake's circumnavigation, 1580.

Georg Braun,
1574

THE PLAY OF INTELLECT IN the creation of sixteenth-century maps worked on two levels and had two phases. In the earlier phase it was the very shape of the world that was re-fashioned through graphic experiments, experiments which made use of science certainly, but whose origin was also highly intuitive. World images such as those of Rosselli, Dürer or Rotz offered new visual perceptions of an expanded world. Perspective and geometry, tools of the Renaissance imagination, were used to articulate a new sense of measured space within which the world map existed. In the second and somewhat later phase, in maps such as that by Apian, the Fool's Cap, and this map by Georg Braun, the play is upon the setting of the world map, its visual context, and the symbolic keys which are provided to interpret the image. Indeed in some cases the map itself is used as a visual symbol within a wider statement about man and his environment.

It would be difficult to conceive of a starker image of worldly political power than this world map, transubstantiated into the very flesh of the Holy Roman Imperial eagle. The allegory of the emperor's mastery over the world is reinforced by the perfect symmetry of the design: the sense of equilibrium may be taken to represent peace, strength, justice or any of the qualities that flow from good government. The specific occasion or circumstances of the map's publication are not known. The Habsburg Emperor, Maximilian II (1564–76), was a humanist and patron of the arts whose liberal policies gave an interlude of peace between Germany's religious conflicts. He campaigned, unsuccessfully, against the Turks on his eastern borders, and extended freedom of worship through much of his realm. Although the eagle is clearly not a symbol of peace, it conveys in this context the sense of benificence achieved through powers justly used. On the wing feathers are painted the armorial shields of the principal cities and rulers subject to the Emperor. In the borders the seven planets, the sun and moon, and the zodiac symbols surround the earth.

The practice of dedicating maps to the rich and powerful originated among engravers and publishers in the sixteenth century in the search for reward and patronage. This image clearly operates on a very different level from the mere dedication, portraying through its structure the Emperor as the force presiding over the whole world. This is a theme which was to re-appear many times during the seventeenth century, though rarely in so dramatic a form. It can be seen as a secularized version of the medieval map structure, where the world waits under the gaze of Christ Pantocrater. And ultimately it is difficult to resist the comparison with the cosmological charts of the ancient near east, where the overarching figure of a god or goddess dominates the world. The association of the world map with power, spiritual or secular, creative or sustaining, is a deeply-rooted and recurring image.

Fool's Cap World, c.1590

THIS STARTLING AND DISTURBING IMAGE is one of the enigmas of cartographic history. The artist, date and place of publication are all unknown, and its purpose can only be guessed at. The geography of the map closely resembles the world maps of Ortelius published in the 1580s, giving a tentative date of c.1590. It is the earliest known use of the world map in a visual joke, although other world maps of this time do contain elements of *jeu d'esprit*. Its central visual metaphor is the universality of human folly, and the various mottoes around the map serve to reinforce that theme. The panel on the left bears the key text:

> 'Democritus laughed at it, (i.e. the world)
> Heraclitus wept over it,
> Epichthonius Cosmopolites portrayed it.'

Thus Epichthonius Cosmopolites is the author's or artist's assumed name, but since it translates roughly as Everyman, his true identity is still hidden. Democritus and Heraclitus were characterized in classical literature as respectively the laughing philosopher and the weeping philosopher. There is a clear reference to this engraving in Burton's *Anatomy of Melancholy* published in 1621: 'All the world is mad, is melancholy, is (which Epichthonius Cosmopolites expressed not many years since in a map) made like a fool's head.' With the exception of this reference, the Fool's Cap map has defied all the efforts of modern researchers to establish its author or its context.

There is an extensive legacy of literature and popular art on the theme of the Fool from the fourteenth to the seventeenth century, and many studies have been made of his social and artistic role, in relation to Shakespeare for example. *The Ship of Fools* was an immensely popular satirical work of the sixteenth century, translated and read all over Europe. Nevertheless the Fool remains an emblem of the late medieval world that is to us enigmatic and alien. The Fool's origin and central role seems to have been in magic: he was a kind of scapegoat who drew down upon himself the forces of evil, unreason or ill-fortune, and by confronting them, averted their power from his community. He was licenced to break rules, speak painful truths, and mock at power and pretension, and the grotesque shape he bore was a kind of living punishment. This frame of reference would have been quite familiar to the audience for this engraving in the 1590's, and they would have recognized in this map a radical visual interpretation of the Fool's role: it is the now the whole world which takes on the Fool's costume, thus forcing the viewer to confront the possibility that the whole created order is irrational, alien and threatening.

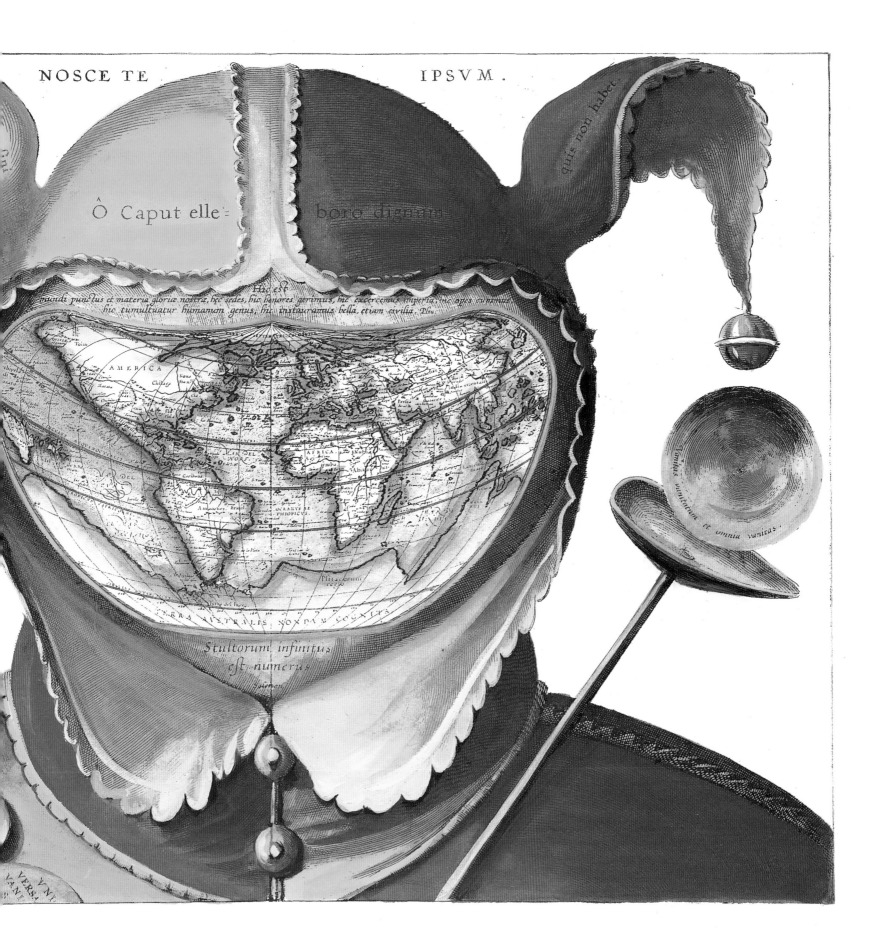

Van Den Keere, 1611

THIS WORLD MAP BY PIETER van den Keere is perhaps the archetypal world image from the age of maritime war, trade and science. The map incorporates some very recent geographical discoveries, as it had to in the intensely competitive field of Amsterdam publishing. A voyage to Novya Zemlya in 1610 is described, and its failure to establish its status as island or promontory. The most noticeable defects of the map are those of the science of the period: the inability to calculate longitude accurately resulted in the east-west extensions of the Mediterranean and of North and South America which are shared by all maps of this era. The map is very large, engraved on twelve separate sheets, to be joined and wall-mounted. It was in fact to be used as a piece of artistic furniture, and the map itself is merely the starting-point of the design: surrounding it is a framework of geographical imagery which raises the map to the level of theatrical spectacle.

Drawing from all four continents, fourteen cities are portrayed, eighteen pairs of representative figures, and the rulers of seven nations, including China, Turkey, and the Holy Roman Emperor. No doubt these portraits are stylized, at several removes from authenticity, but it is the geographical scope which is important. These are monarchs who control the world's destiny, and these are the cities they rule over around the world. Inside this framework are the six symbolic female figures of the continents, with Europe at the centre as queen of the world receiving the fruits of art, war and trade. The new world is represented by three figures, Mexicana, Peruviana and the giantess Magellanica (South America). No less striking are the three decorative tableaux on the map itself. The cartouche in North America shows Vespucci, Magellan and Columbus handling instruments of navigation. In the Pacific Ocean, the largest allegorical tableau almost certainly alludes to the truce negotiated between Spain and the Netherlands in 1609: the figure of Peace is crowned by Victory, while Justice holds War in chains. In the eastern hemisphere an Academy of ancient and modern scientists — Euclid, Archimedes, Ptolemy, Mercator, Ortelius and Tycho — are gathered around a celestial globe under the guidance of the figure of Astronomy. To their right is an elaborate *vanitas*, a man whose life and work is threatened by the figure of death. Ornament outweighs substance, and ultimately the map appears to be like a ship foundering under the weight of its exotic cargo. The modelling of the figures, the visual depth in the tableaux and the ingenuity of the details all suggest that the artist had learned from the contemporary works of Rubens and his school.

This map is effectively a paper theatre of the world, enabling the Amsterdam merchant to contemplate at his leisure the kings, the cities and the peoples of the earth. Like the medieval *mappa mundi*, it is a cultural icon, an image of the world interpreted according to the knowledge and values of its time.

(*This map has been coloured here in the style of the period.*)

APPA, EX OPTIMIS AUCTORIBUS DESUMTA, ſtudio Petri Kerl.

Ottavio Pisani,
1612

By the seventeenth century there was an almost universal satisfaction, intellectual and aesthetic, with the twin-hemisphere map. But in that same period, progress in geometry, astronomy and optics was creating potentially a new theoretical basis for mapping, and the seventeenth and eighteenth centuries witnessed some surprising experiments in theoretical cartography. Of these one of the most capricious is this polar projection by Ottavio Pisani, where the entire globe is compressed into a single sphere, centred most unusually on the south pole. Its most startling feature is that all the land-masses are engraved in reversed form, as mirror-images of their true shape, although the names read correctly. Why has this been done, and what was the author attempting to demonstrate?

Unfortunately there is no external evidence with which to place Pisani's map in a precise context. Pisani is such an enigmatic figure that one is tempted to believe that the various accounts of him may have been conflated from two or more people of the same name. A Neapolitan, he published some devotional verse in Rome in 1603, then enlisted in the Spanish army in the Netherlands, attaining high rank. Settled in Antwerp, he achieved some fame as an astronomer, and published in 1613 a treatise on stellar motion, and apparently a discourse on the philosophy of law. The date of his death is unknown. There is every indication that this work was intended not as a functional map, but as an optical, mathematical exercise. In the first place, the lands around the sphere's edge are grotesquely distorted because on this projection the linear scale increases progressively from the centre outwards. In the second place, the chosen point of projection, the south pole, was at this time the most uncharted region on earth. Pisani has clearly chosen this structure not for any practical purpose, but to demonstrate a phenomenon which he was the first to notice: that a world map projected on a plane tangent to the south pole will reverse east and west, and produce a mirror-image map. This occurs because the south polar projection effectively turns the globe upside down, so that longitudinal measurements from the fixed central meridian reverse the left or right orientation that would be normal from the north. In this case, Pisani had adopted the Ptolemaic central meridian (bisecting Arabia), and plotted west to the right of that line and east to the left, as had all mapmakers in the northern hemisphere. But on this basis, the drawing for example of the coast of South America from Cape Horn to Brazil, a north-easterly bearing, appears to trend north-west, producing the mirror-image effect. To counter this, a conscious decision would have to be made to reverse the natural east and west on the graticule outline before the map is plotted.

What Pisani is really pointing out is that the all-important 'cardinal' directions on the sphere become relative when the sphere is transferred to paper. The conceptual or interpreted world strives for an objective viewpoint, unrelated to the location of any human observer. Mathematical or optical manipulation can be used to create world images that are aesthetically satisfying, or to undermine them through logical re-modelling.

Pisani obviously became aware of the mystification his map had caused. A new edition was published in 1637 in which the mirror-image was rectified. That map includes a small diagram explaining the mirror-image principle, and more remarkably introduces the personality of Pisani himself. He is shown seated at his drawing desk, plotting from globe to paper. His armour is now laid aside, a defiant stare is on his face, and above him are the words 'To my detractors: Oh you who criticise my work, You do better: Try to make a globe both round and whole, drawn on a flat plane like this'. There is surely no other example of a map publisher confronting his public in this way, taunting them to do better if they find fault with his maps. Perhaps it is not surprising that Pisani's career as map publisher was limited to this one essay. His theoretical approach had its roots outside of the world of professional mapmaking, and had no impact upon it. It stands as an almost cerebral reminder that despite the conventions of map publishing, the sphere is always a sphere, irreconcilable to the paper world.

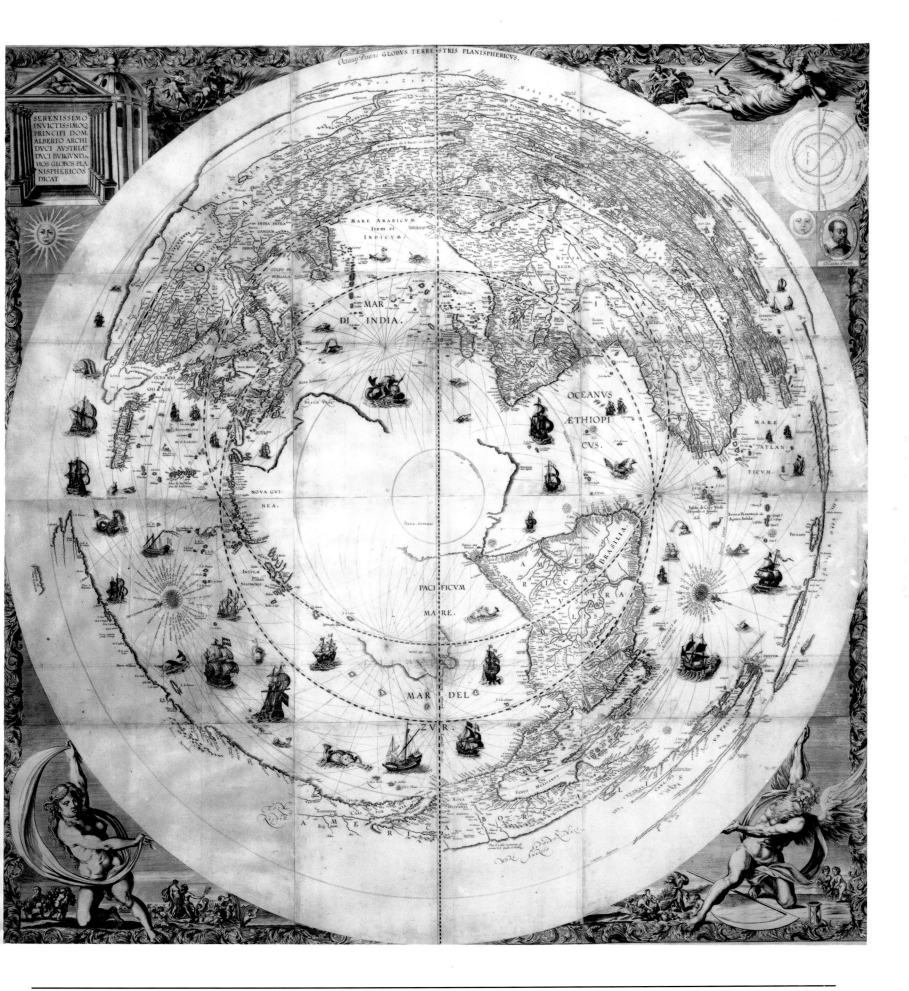

OTTAVIO PISANI, 1612

Franc-Antoine de la Porte, 1617

W<small>HEN</small> M<small>ARIE DE</small> M<small>EDICI,</small> Q<small>UEEN</small> Mother to the young King Louis XIII, was meditating the building of the Palais de Luxembourg and its gardens, she commissioned the great architect of the day, Salamon la Brosse, who in turn was involved with scores of lesser designers and artisans in planning the innumerable features of the house. Among those received was an ingenious design submitted by Franc-Antoine de la Porte for a formal garden laid out as a world map. The map itself is unremarkable and there is no record that it was built, but the very concept provides a forceful insight into the contemporary perception of the world map.

The dedicatory verse over the gate may be rendered thus:

'Since Eve first through her guilty hands
Did change her dainty paradise for thorns,
Marie the Royal Matron now transforms
To paradise this earth and all its lands.'

The Italian gardens which were so admired in the seventeenth century sprang from the dramatic possibilities inherent in the sites; but terraces and waterfalls did not transfer well to the plains of northern France. Instead, extended planting over vast areas was used to create the grandest possible domain. Sub-divisions into geometric compartments were designed with trimmed hedges of herbs and shrubs, which in this case would form the continental outlines. Coloured stones and sand were used as often as flowers to complete the ground pattern, here the place-names, equator, tropics and other details.

The metaphor of the world as a theatre of human endeavour is very familiar in late Renaissance literature. There could be no clearer image of secular ambition than the creation of a parallel world in microcosm for the delight of the royal patron. That the recipient of this votive fantasy was a woman, has introduced a further level to the symbolism: by the exercise of good government, Marie de Medici can now undo Eve's fall, and revive Eden. This sense of freedom in manipulating the world image within political or artistic metaphors, is entirely characteristic of the years 1550 to 1650. Its origin lies in the Renaissance impulse to celebrate the world in its secular, theatrical, or governmental aspects. It is an implicitly competitive, rapacious, even brutal view of the world, but in this case that resonance is softened through the appealing metaphor of the garden-world.

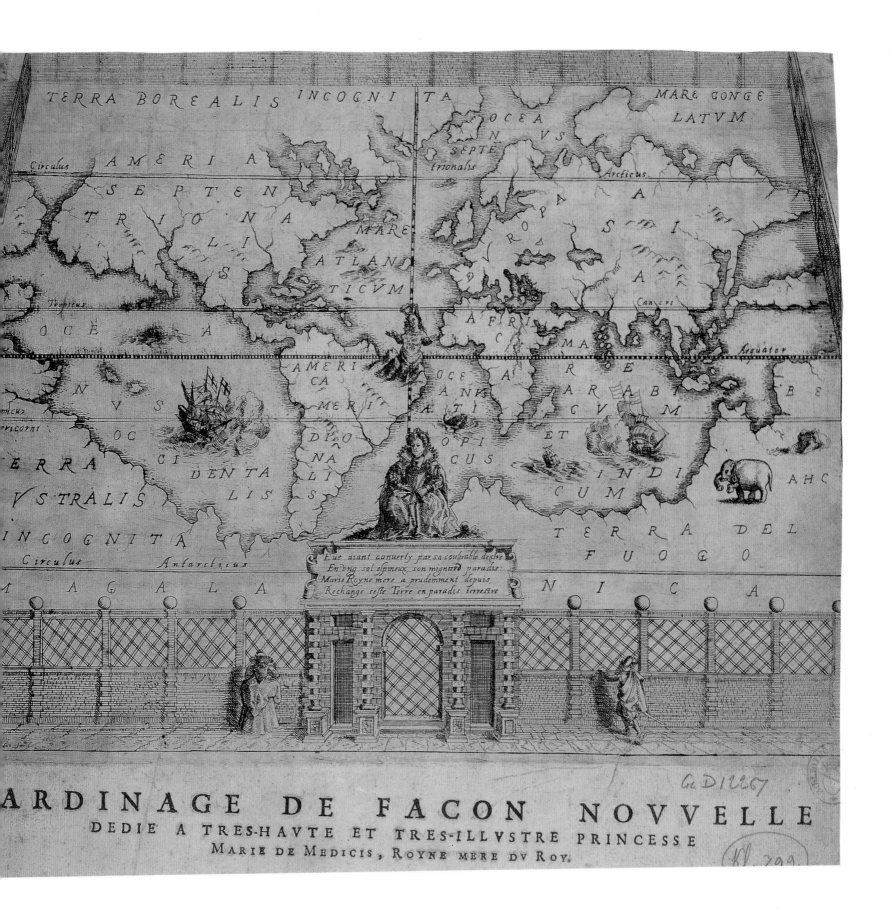

TERRA BOREALIS INCOGNITA   MARE CONGE LATUM

OCEANVS SEPTEntrionalis

Circulus   AMERIA   EVROPA   S   Arcticus

SEPTENTRIONALIS

MARE ATLANTICVM   Cancri

Tropicus

OCEA   AFRICA   MA   RE   Æquator

AMERICA   OCEANVS   ARABICVM   BE

MERI   ÆTHIOPICVS   ET

DIONALIS   OPICVS   INDICUM   AHC

OCCIDENTALIS   TERRA DEL FUOGO

VSTRALIS

INCOGNITA   Circulus Antarcticus   NICA

MAGALA   LA

Eue aiant conuerty par sa coulpable dextre
En vng sol espineux, son mignard paradis
Marie Royne mere a prudemment depuis,
Rechange ceste Terre en paradis terrestre

ARDINAGE DE FACON NOVVELLE
DEDIE A TRES-HAVTE ET TRES-ILLVSTRE PRINCESSE
MARIE DE MEDICIS, ROYNE MERE DV ROY.

Antonio Sanches,
1623

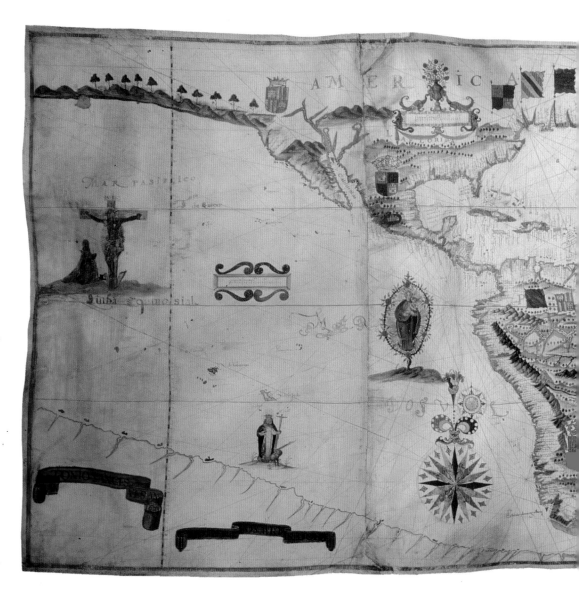

ONE OF THE WAYS IN which map history relates most clearly to art history is through the phenomenon of national schools of map-making, and the way the centre of conceptual and stylistic innovation has periodically shifted from place to place. During the flood-tide of the age of discovery, Spain and Portugal were naturally the centres where the latest geographical knowledge was received, interpreted and embodied in new maps, and a corpus of manuscript charts resulted, constructed with great empirical skill and artistic boldness. Yet this manuscript tradition was not accompanied or succeeded by the development of map printing. Elaborate sea charts and world maps continued to be painted by hand in those countries for more than a century after Europe had embraced the new print technology. The culture of secrecy in these absolutist regimes prevented the new medium from being applied to the potentially sensitive area of geography, although literary printing became well established. Portuguese seaborne power in the sixteenth century created a strong trading monopoly rather than an empire, spread along the coasts of Africa, South Asia, and South America. To the motive of trade was added that of religion: missionaries, especially Jesuits, carried the Catholic faith wherever the explorers and conquerors led.

Portugal's place at the centre-stage of world affairs inevitably ended after Spain annexed the throne of Portugal in 1580, and Spain's European enemies attacked her overseas. This map is a splendid if sombre memorial to Portugal's Golden Age. Detailed and accurate in its delineation of the Mediterranean, Africa and South Asia, its plane-chart construction renders it a highly distorted and anachronistic image of the world: the science of Renaissance map construction, from Ptolemy through Waldseemüller, Rosselli, and Apian to Mercator, does not exist for Antonio Sanches. The arms of Portugal and Spain dominate the world, while the continental interiors are filled with bright meaningless shapes, so that the map resembles a naive tapestry. The dominant imagery of the map is religious: a crucifixion, a madonna and child, St. Anthony of Padua (Portugal's patron saint), and Franciscan and Dominican missionaries survey the world's oceans. A glowing cross commemorates St. Francis Xavier, first Christian missionary to Japan. The purpose of the map is almost to chart the spread of Christianity through the new worlds. At a time when the leading Dutch mapmakers were developing an array of classical and secular imagery to decorate their maps and to express their nation's mastery of the now encompassed world, the Portuguese chartmaker has tried to affirm that the essential spirit and legacy of his nation's worldwide empire was religious. This image summarizes powerfully the cultural isolation of the Iberian manuscript map at this late period, as Portugal entered the twilight of her international power

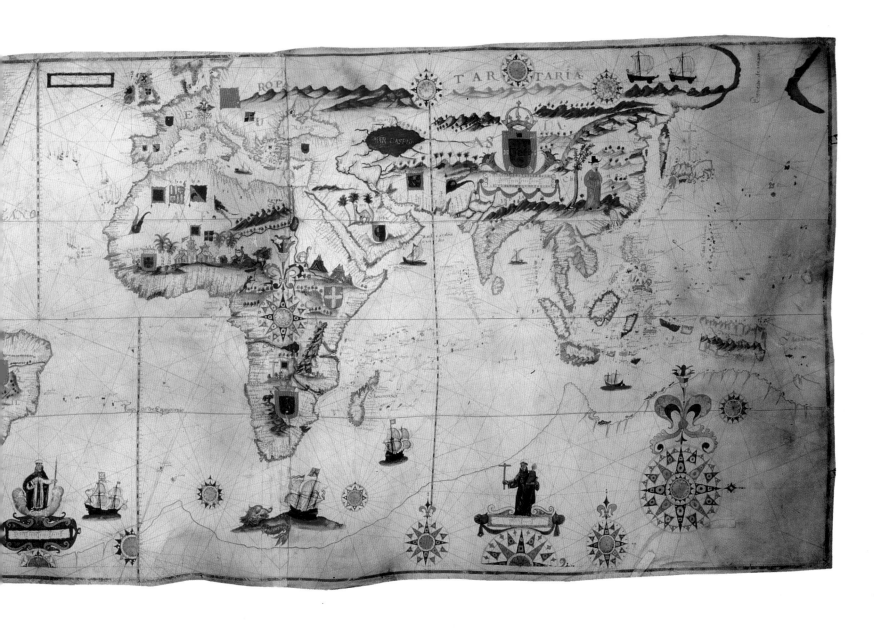

Samuel Purchas, 1624

THE INITIAL MOTIVE OF THE early voyages of discovery was of course the search for a sea-route to the east. The possibility of colonization in the newly-discovered continents dawned slowly on the European mind, and not until the late sixteenth century did France, England and Holland consider challenging the Spanish presence in the Americas. In England the writings of Richard Hakluyt in the 1590s were instrumental in spreading colonial ideas and ambitions. His great book *The Principal Navigations, Voyages and Discoveries of the English Nation* (1589 and subsequent editions) is a key historical source for our understanding of the events and the motives of that period. After Hakluyt's death in 1616, his work was continued by Samuel Purchas, who published in 1624 his own travel and exploration epic *Hakluytus Posthumus, or Purchas His Pilgrims*.

The title-page of this work is an elaborate icon of the Jacobean world-view. Most of the apparatus — historical, literary, political — which is normally found on the seventeenth-century map itself, has been externalized so that the title-page resembles a Jacobean facade, designed with numerous stories, panels and niches into a rather disorderly architectural whole. The small world map is an unusual type of twin-hemisphere, centred in the Pacific and Atlantic oceans, its model clearly the Hondius-Drake map of 1589. On the map appear the words 'Soldiers and Merchants, the world's two eyes to see itself'. This curious phrase is an ingenious conceit: on the one hand it is a perceptive summary of the colonial spirit, while simultaneously pointing to the visual structure of the double-hemisphere map. This metaphysical and typically Jacobean delight in word play dominates the book's almost interminable title, which includes the words 'A world of the world's rarities are by a world of eyewitness authors related to the world'.

The nationalist fervour of Purchas's geography is clear from the memorial to Elizabeth I, and from the igenious double-portrait of King James I and Prince Charles astride a map of Britain. Purchas, like Hakluyt, had a triumphalist vision of England's role in the contemporary world, and he places that belief within a long historical perspective by means of the crowded portrait gallery that dominates the page. From Noah, the first shipbuilder, we see a procession of the world's explorers and conquerors, in the worldly and spiritual senses: Abraham, Ulysses, Alexander, Caesar, St. Paul, Tamberlaine, Constantine, Columbus, Magellan and many more. As Purchas comes nearer to his own period, the gallery becomes more English, with Richard the Lionheart, Edward I, Sir John Mandeville, Drake, Cavendish, Davis and Hakluyt himself. This cast gives a vital insight into the nature of the seventeenth-century world-view as a vast theatre of human endeavour and achievement. Driven by religious or worldly energy, intellectual curiosity, or physical courage, characters as diverse as Tamberlaine and St. Paul have played their parts in re-shaping man's world-view. England's part in this drama is carried back to the crusading king Richard (or, in the realm of legend, to Joseph of Arimathea who brought the Holy Grail to England), and the moment for England to assume the leading role in exploring and settling the new world, has now arrived.

Purchas's title-page is a conceptual chart of the seventeenth-century world seen through English eyes. Its cultural apparatus provided a coherent basis for Jacobean thought and action, and the vision of the world dominated by 'soldiers and merchants' was to prove prophetic.

Bankoku-Sozu, 1645

THE TRADITIONAL WORLD MAPS IN Japan before the era of European contact were religious and symbolic rather than representational. The most prevalent type, and the one of which we have the clearest knowledge, was Buddhist, and had the formal title *Gotenjiku-Zu* — 'Map of the Five Indies'. It centred on India rather than Japan, because of its Buddhist origin, and although certain features such as mountains and rivers were clearly shown, it was conceptual rather than geographical. The arrival of Jesuit missionaries in Japan in the 1550s introduced a totally new understanding of world geography. This process was perhaps inevitable, although the western world map came not directly from St. Francis Xavier and the first Europeans, but via China. The Italian Jesuit missionary Matteo Ricci arrived in China in 1582, and, perceiving the advanced level of culture in that country, he determined upon a missionary strategy based on the concept of cultural exchange. He and his colleagues learned Chinese and studied Chinese science and literature in order to secure acceptance as equals in their adopted society. Simultaneously they translated European works into Chinese and publicized European concepts, not only Christian apologetics, but works of science, philosophy and history. In this context, a western world map was a fundamental teaching document, and in 1584 Ricci supervised the woodblock printing of a large world map in Chinese, drawn on the prevalent oval projection. This was to be reprinted several times, thousands of copies being reportedly sold to an eager market. In his *History of the Introduction of Christianity to China,* Ricci comments proudly 'The Fathers gave such clear and lucid explanations on all these matters which were so new to the Chinese, that many were unable to deny the truth of all that they said; and for this reason the information on this matter quickly spread among all the scholars of China. From

this one can understand how much esteem was given to the Jesuits, as well as to our land, which thenceforth they did not dare to describe as barbarian, a word they were accustomed to use in describing countries other than China.'

The cultural impact of Ricci's map was profound, and at this period Chinese science and scholarship was the major influence on Japanese intellectual life, so that a Japanese version of the map established itself as swiftly as the original had in China. This edition, printed from woodblocks in Nagasaki in 1645, was accompanied by a print of forty pictures of the peoples of the world; indeed the title *Bankoku-sozu* is a generic one, meaning simply 'Complete Map of the Peoples of the World'. The principal visual and cultural departure from the European model is that it is Pacific-centred. The appearance of this map and the position it swiftly

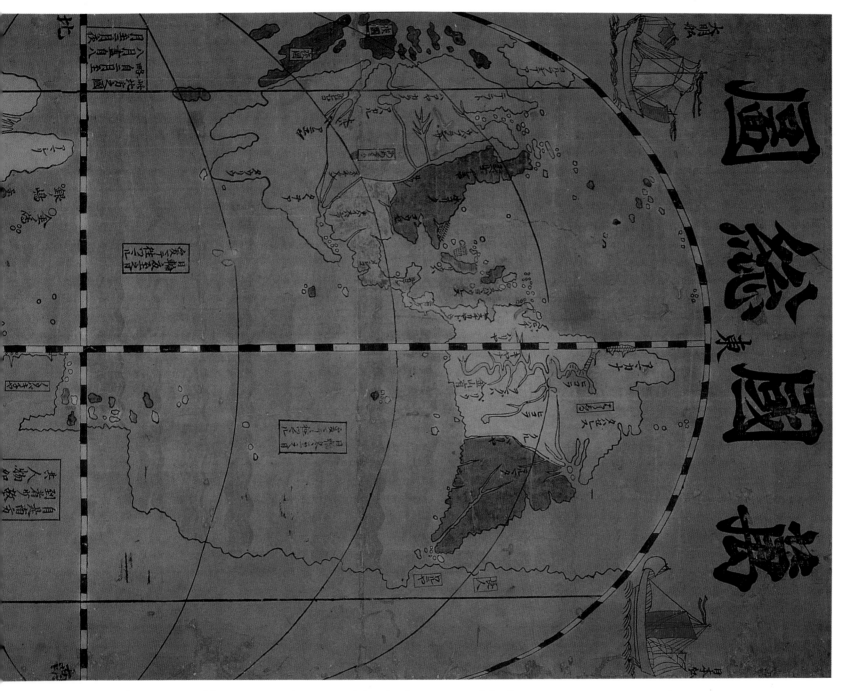

assumed as the dominant world-image is perhaps surprizing in the context of the Japanese policy of *Sakoku* — Seclusion — implemented in the 1630s, which endured for more than two centuries. However the policy of seclusion sprang apparently from motives of internal political alliances and enmities, and seems not to have been a xenophobic cultural hostility. Nevertheless the Jesuits were expelled, Christianity ruthlessly eliminated, and European traders barred. It is significant that this map was produced in Nagasaki, since that was Japan's sole point of international contact, by special licence granted to the Dutch. The seed of the European geographic concepts once planted, flourished despite this isolation, in a sequence of Japanese world maps, unchanged for over two centuries, a remarkable example of instant but enduring cultural assimilation.

Pseudo-Blaeu World, 1665, republished 1683

Eᴀʀʟʏ ɪɴ ᴛʜᴇ ꜱᴇᴠᴇɴᴛᴇᴇɴᴛʜ ᴄᴇɴᴛᴜʀʏ the exuberant imagery of the twin-hemisphere map undoubtedly added a fresh dimension to the world map, reviving the visual-encylopedia aspect of maps so strongly evident in the medieval *mappae mundi*. But given the publishing practices of the seventeenth century, these motifs were mercilessly copied and re-copied until they became like exhausted poetic clichés, stripped of any contemporary resonance: the eye glides over them without effort or response.

This map appeared under Blaeu's name and is clearly a flamboyant celebration of the mapmaker's art; but the richness of the twin-hemisphere map has descended here into ornate heartlessness. The claimed authorship is quite spurious: Willem Blaeu died in 1638, while this map bears a dedication dated 1665 (to Jacopo Altovito, papal legate in Venice), and Blaeu's name is stolen purely to lend authority to the work. In fact this map has been directly copied from an earlier model published in Amsterdam in 1629 by a little-known scholar, Nicolaes van Wassenaer, and the anonymous publisher of this version has made no attempt to update the geography. The style of the map itself is anachronistic in its use of sea-creatures and compass roses, while the mesh of compass-lines is quite without navigational sense. The border scenes have been crudely engraved, robbing them of their intended impact: the contrasted scenes of the age of gold and the age of iron, the four elements, and the portraits of the navigators are slovenly and without individuality. The whole impression is of a work that is derivative in conception and debased in execution. It marks the decline of an artistic form which had outlived its age, and was soon to be revived with new ideas and new themes.

This is not merely a matter of execution but of intellectual context. It should be born in mind that the revolutionary decades of the Age of Discovery belonged only to Spain and Portugal. After Magellan's first circumnavigation in 1522, half a century would elapse before Britain, France and Holland assumed their roles in search of conquests and empires, and a new self-consciously heroic phase of overseas voyages began. It was from the 1580s onwards, the years also when the great Flemish mapmakers began to issue their newly-conceived world atlases, that the world map regained its encyclopedic, celebratory form. To this age heroic, classical imagery was a natural idiom for the new world map. But by the final quarter of the seventeenth century, the realities of trade and international rivalry, and the spirit of science, had made that idiom inauthentic and dated, and its appearance in this map is purely rhetorical.

The decorative feature which is of contemporary interest is the panel of equestrian portraits which is not an integral part of the map at all. The kings of France, Spain, England and Poland are shown, together with the German emperor and the sultan of Turkey, all flanking a portrait of Pope Innocent XI. The juxtaposition of these contemporary rulers provides a clue to this map's re-publication by an anonymous, probably Italian, engraver. In 1683 when the Turkish siege of Vienna appeared as a grave threat to all Christian Europe, Innocent XI inspired and financed the counter-attack led by John III of Poland and the Emperor Leopold. It is highly unlikely that portraits of the four principal actors in this drama should appear together by coincidence, and this special edition of the map was almost certainly published to commemorate the end of the siege of Vienna.

ARUM SIVE NOVI ORBIS TABVLA, auct: G. Blaeu

Frederick de Wit,
1668

THE BAROQUE TWIN-HEMISPHERE WORLD map lost its integrity because its form became divorced from its content. Early in the seventeenth century, the Dutch in particular were justified perhaps in sensing the heroic element in maritime exploration and trade, and in creating a highly-wrought artistic idiom with which to celebrate it. But with the rise of the scientific spirit later in the century, the world map would inevitably be compelled to assume a more serious and scientific form, and outgrow the rhetorical gestures of the past: the divorce between form and content would bè healed by the creation of a new cartographic idiom rooted in science. Before that occurred, the later seventeenth century produced many purely rhetorical versions of the baroque twin-hemisphere map, but also some elegant attempts to solve the dichotomy between form and content.

In this publication, a copy of a 1665 original, the cartographer begins by cleansing the map itself: there are no ships, no sea-serpents, no compasses, no rambling texts, no writhing demigods. The geography is equally restrained: the known coasts of Australia and New Zealand (discovered by Tasman in 1642–44) are given a minimalist treatment; the bloated north-west of America is replaced by unexplored ocean, and the conjectural polar continents are dissolved, the south polar inset almost a challenge to the geographic theorizers. Having rationalized the map, the cartographer concentrates all his artistic energy into the four borders. Their subject — the four elements — is perfectly conventional, but they are handled here with a virtuosity that is outstanding. The elements are not personified into single classical figures alone, but are dramatised in complex, highly fluid scenes. The element of fire is presented through an image of

war, and the figure of the three-headed dog Cerberus, guardian of Hades, suggests the release of destructive forces into the world. Earth is symbolized by the harvest, and the hunstmen's return, while a cannon rusts on the hillside, relic of a forgotten war. Graceful Dutch ships drift on a calm sea, while above the natural forces of wind, rain and lightning, the gods of the air act out their timeless parts against the background of the revolving zodiac.

Each of these tableaux is a self-contained miniature landscape, employing the techniques of perspective, with elements of the composition skilfully placed in the foreground, the middle-ground and the background. They were etched by Romein de Hooghe, perhaps the most daring and accomplished of the Dutch engravers of this period, whose work has a grace and a vigour that is instantly recognisable. But the ingenuity and clarity of these images are ultimately sterile, since there is clearly no conceptual relationship between them and the map itself. The virtuosity of the art merely serves to emphasise the division between the form and the content. These allegorical motifs are of extreme antiquity, their ultimate source probably the description in *The Iliad* of the shield of Achilles on which the elements of human life were displayed — peace, war, harvest, the land, the city, the dance, the hunt, all surrounded by the ocean. To be modelled on classical antiquity was one of the canons of taste in late seventeenth-century art, but the use of these motifs to frame the rationalized world map was a dislocation of sensibility. It represented an account with the forces of reality that was becoming seriously overdrawn, and which would have to be squared if the world map was to reflect accurately the culture from which it sprang.

Nova Totius
TERRARUM
ORBIS
TABULA,
ex officina F. de Wit.
Amstelodami

CHAPTER FIVE

Science and Communication

'OF THE INTELLECT OF EUROPE in general the main effort, for now many years, has been a *critical* effort; the endeavours in all branches of knowledge — theology, philosophy, history, art, science — to see the object *as in itself it really is*.' Matthew Arnold's perception expresses forcefully the major trend of intellectual history since the later seventeenth century. As knowledge became self-conscious, the disciplines of knowledge perceived an imperative demand to transcend subjective, cultural or non-rational forms of thought, and to address clear-sightedly the object, the thing itself. Whether such a programme could ever be achievable in any discipline is open to question; it fails to do justice to the cultural context in which all knowledge exists, and the different levels on which ideas are expressed and interpreted. In the evolution of the world map, the overwhelming fact in the eighteenth and nineteenth centuries is that maps became 'more scientific'. But all this really means is that in their construction the maps conformed to increasingly clear and self-conscious technical rules. They did not thereby cease to express the cultural concerns of those who produced and used the maps. The acceleration of social change in the modern period has taken the world map out of the confines of geography into the realms of science, popular culture and communication. Advances in literacy, wealth and mobility created new markets for maps and new means of producing them: as secularization was the outstanding feature of world maps in the Renaissance period, so democratization is the key process since the Enlightenment — democratization of both the contents of the map and of the map as an object.

Few shifts in intellectual taste have left their mark so clearly upon the world map as the scientific revolution of the later seventeenth century. Within the two decades 1680–1700, the classical and heroic integuments of the world map were shed, and a new repertoire of scientific motifs became the vogue. The centrality of cosmology in eighteenth-century science led to the dominance of astronomical motifs and texts, setting the image of the world effectively within its natural context in the solar system. The new science entirely re-shaped the intellectual landscape: theologians and scientists agreed on the manifest harmony between revelation and reason, and Leibniz notoriously crystallized the contemporary creed of optimism by arguing that since God had created the world, it must be ruled by harmony and order. Science was in vogue, and every gentleman's study boasted its globe, its telescope, and its volumes of the Royal Society's *Philosophical Transactions*. Maps of the world and the heavens even appeared engraved on snuff-boxes. The twin-hemisphere form continued to dominate, the inertia of commercial map-publishers making them unwilling to experiment, although the Mercator chart made something of a comeback as its value at sea

came to be universally recognized. As the eighteenth century progressed, large atlases from leading publishers commonly included both a twin-hemisphere world map and a Mercator chart.

The Mercator chart was also found to be suited to the other type of scientific world map which now emerged — the thematic map. Pre-figured once or twice during the eighteenth century, the thematic world map really came into its own in the nineteenth century, with the explosion of scientific and social data on an enormous range of subjects. The charting of the invisible, which is precisely what thematic mapping attempts to do, marks a decisive shift in the perception of the world map. The map is clearly being raised from its basis in direct geographic representation, and used as a model or analogue to demonstrate conceptual categories of reality. Thematic maps appeared on agriculture, biology, climate, commerce, disease, ethnology, geology, religions, languages, military power and a myriad of other subjects from around 1820 onwards. Many of these maps were of course polemical, whether consciously or not, seeking to reinforce a selective view of the world. At the same time the exciting new technique of chromolithography was used to good effect to revive the illustrated world map, with vivid images drawn from travel and adventure, trade and empire, and the impact of this type of world mapping was vastly increased by the new technology which produced it. When maps were engraved on copper plates, and each copy individually pulled by hand, one thousand copies was a large edition of a printed map. With the advent of the power litho press, tens of thousands became feasible. Mass production had implications for both the content of the maps and their marketing, placing them firmly within the realm of popular culture. Such maps helped to shape popular perceptions of the world, and popular cultural and political values. Arnold's concept of seeing the object as it really is cannot be sustained from an analysis of the cultural forces acting on the world map. The diversity of world images that came with the new scientific perceptions and the wider market, rendered the ideal of objectivity a naive one. Nor could the established intellectual and craft monopolies of map production exercised by the publishers of Amsterdam, Paris and London, survive changes in technology and literacy: by the mid-nineteenth century world maps were appearing in Berlin, Boston, Calcutta, Edinburgh, Madrid, Philadelphia, New York, Stockholm and Tokyo.

In the twentieth century sophisticated scientific techniques have, paradoxically, further dissolved the simple ideal of objectivity in the face of conceptual diversity. Thematic mapping has been extended to show the invisible in a truly radical sense: vast geological landforms beneath the earth's surface, the shape of the continents millions of years ago, the floor of the oceans — in all these the scientific

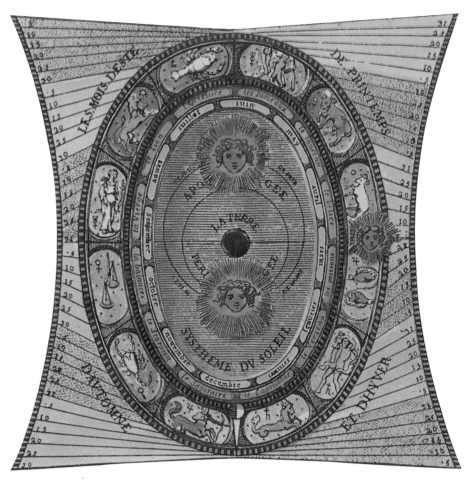

In Jaugeon's world map of 1688, the sun's human face is a tribute to Louis XIV as the Sun King, the central focus of the entire world map.

Mapseller's Shop by Guillaume Delisle: the commercial map trade around 1700.

imagination has reconstructed images of the world which the human eye will never see, objectively true though they may be. The science of map projection serves likewise to dissolve the certainties that once accompanied the Mercator or the twin-hemisphere world map. The expanding number of possible map projections and their claims to pysychological or political correctness, show the intellect toying with the impossible.

Modern technology has achieved what is usually regarded as the most decisive step in the history of world mapping. In the period of more than 2,000 years since the earliest world maps were made, ours is the first generation to escape from the surface of the earth and see it as an object, a planet in space, a sphere composed of land, water, ice and cloud. The visual reality that had always been beyond human reach has become a commonplace through the satellite image and the space photograph. This breakthrough in perception has come to be symbolized by a contemporary icon: the multicoloured circle of the earth seen against the star-filled darkness of space. But how objective, how complete, how informative is this image, with its god-like perspective on our planet? Photographs of the entire earth reveal a small violet disk surrounded by swirling cloud, beneath which the continental outlines are just discernible. As an image its power derives from our prior knowledge of what it is and how it was obtained; but it is scarcely enlightening, concealing much more than it reveals. Compared to a map, which is interpreted, conceptual and diagrammatic, it resembles an incomplete or damaged mosaic. For more precise definition we turn to larger-scale satellite imagery. But here we encounter a paradox: satellite data must be extensively processed to build meaningful images from it. Colour and planimet-

ric form have both to be manipulated to transform digital information into visual form. A satellite image of Sicily or Skye or Massachusetts is no more the objective reality of those places than a conventional map is. And this is interpretation at the level of origination: interpretation of the image by the user must follow, understanding what the false colours signify and how the various features, natural or man-made, may be recognized. We understand and appreciate satellite images because conventional maps have taught us what to look for. It is significant that satellite images are in fact increasingly combined with structural and conceptual elements from conventional maps so that they may be interpreted: the language of cartography, evolved over five centuries, has proved durable and flexible enough to interpret this new generation of source data undreamed of by earlier mapmakers.

Today we live in a world of images, and we regard it as natural that they should be used to persuade or disturb as well as to enlighten. In contemporary media the world map is a recurring motif, sometimes retaining its symbolic attributes of power and possession, but more often now having an entirely new resonance: that of uniqueness and vulnerability. The age of ecology has added this new perspective to the world map, so that satellite image mosaics dramatize the human impact on temperature, climate, vegetation etc. Thus the data of modern science is channelled by cultural forces to express contemporary concerns. As with the medieval *mappa mundi* or the seventeenth-century baroque theatre, the iconographic dimension of the world map is still dominant. The reality, the object as it really is, eludes us.

The earth from space: a twentieth-century icon.

John Seller, 1673

In the second half of the seventeenth century the double-hemisphere world map was trapped in a cul-de-sac of outworn, conventional imagery. It was rescued and given a further half-century of life by the creation of new framing motifs, drawn not from literature or art but from science. This map by John Seller is from the early transitional phase when the scientific images appeared still alongside traditional motifs.

In the early years of the century there was a certain reluctance among scientists to publish their new findings and theories, when the attitude of secular and ecclesiastical authorities might range from suspicion to persecution. Instead they formed through correspondence 'invisible colleges' for the exchange of ideas. As the intellectual climate improved through the century, these groups became the nuclei of the most important institutional innovation in modern science: the formation of the scientific academies. Of the societies founded in London, Paris, Berlin, Leipzig, Florence and St.Petersburg, the Royal Society in London, chartered in 1662, was the first. Formed with the weight of royal patronage, these societies had the effect of placing science on an equal footing with the traditional spheres of learning, theology, law and the classics. They may be seen as the formal, external marks of recognition of the scientific revolution, of the emergence of science as a self-conscious and significant force in European culture.

The mapmakers' response took two main forms: firstly the map itself became more austere, disdaining decorative and symbolic elements, and secondly their images of the world were set not within an artificial framework of literary images, but within its objective environment of the solar system. In this map, besides the traditional tableaux of the four seasons, Seller has engraved three alternative theories of planetary motion, two lunar charts, and the earth's orbit around the sun. The Copernican theory of a heliocentric solar system was first published in 1543, but it found neither wide nor rapid acceptance. For almost a further two centuries, when planetary systems were displayed, as Seller displays them here, the Copernican model (lower right) continued to be presented alongside the classical system of Ptolemy (lower left) and the 'compromize' theory of Tycho Brahe (upper right). Tycho had accepted that the planets revolved around the sun, but insisted that the earth was still the centre of the system, around which the sun itself revolved.

The most novel feature of this map is the double-chart of the moon, one a natural image, the other a redrawn map naming some of its features. After the invention of the telescope, Kepler had published an early drawing of the moon, but the most detailed lunar maps were those of the Polish astronomer Hevelius, published in 1647. These were the source for Seller's map, and for a lunar globe commissioned by the Royal Society from Christopher Wren, for presentation to King Charles II in 1661. Seller's was the first widely-published moon map in England, and it shows the main features with reasonable accuracy, although the names which he gave to the great seas and craters did not achieve permanence.

Seller's adoption of scientific elements in his world map has not produced any structural change in the global image. The twin-hemisphere design was still regarded by the map trade as the indispensable product for the market, and a further half-century was to elapse before new pressures would revive the Mercator map. But Seller is responding to a discernible shift in intellectual taste, in particular to crucial developments in astronomy, which culminated in the Newtonian vision of space as dynamic, and which had very specific implications for man's image of his planet within the cosmos.

N. Jaugeon,
1688

In seventeenth-century france, as in England, the shift in the intellectual climate towards science was marked outwardly by the founding of an academy, the Académie Royale des Sciences, established with Louis XIV's patronage (and strict control) in 1666, followed soon afterwards by that of the Paris Observatory. The first director of the Observatory was Jean-Dominique Cassini, who pioneered new standards of surveying and mapmaking, and inaugurated in France the first topographic survey of any European country. These developments were to lead in the early eighteenth century to the origin of the science of geodesy — the precise measurement, through complex trigonometry, of the figure of the earth.

The central problem addressed by seventeenth-century science was the mechanism of the earth, the solar system and the universe. When the classical notion of crystalline spheres supporting the heavenly bodies had been rejected, when precise observation had revealed the complexity of planetary motion, and when the use of the telescope had given some conception of the immensity of space, the overwhelming intellectual need was to expose the forces which could hold such a mechanism in balance. Astronomy, or rather stellar mechanics, became for this period the queen of sciences, and the effect on the world map of this intellectual focus was a desire to display the world in its astronomical context. Twin-hemisphere maps now appeared whose borders were filled with models of planetary systems, Ptolemaicœ and Copernican, with diagrams of the phases of the moon, of eclipses and other astronomical motifs.

In this elaborate chart by the otherwise unknown Jaugeon (so unknown that even his Christian name cannot be discovered), the modest world map is surrounded by a myriad of texts and images that reveal the contemporary vogue for astronomy. Each hemisphere is encircled by a dozen texts which form a glossary of geographical terms; subjects such as the zodiac, climates, and the equator are defined at some length, while between the texts the principal constellations are drawn. Along the central axis of the map between the two hemispheres are three highly-wrought diagrams showing the phases of the moon, the mechanism of the seasons and the zodiac, and the three contemporary planetary theories, while the corners of the map show calendar-calculating tables. Jaugeon's title, in the ribbon which flutters across the top of the map, is curious and revealing: *A Set of the World's Sciences, embracing the World, the Heavens, and Civil Life.* The ambitious intellectual claims which this map makes for itself contrast strikingly with the mannered, rather vacuous world maps of the preceding decades. Science has not however banished art and politics from the map. The stucco style of the map's framework is highly unusual, as if it were moulded like a Rococo relief, while the centrality of the sun in the map's structure, and its human face, is undoubtedly a play on the personal emblem of King Louis XIV, the sun king. The image of the world is still, in the age of science, subject to the laws of intellectual taste: this is mapmaking *à la mode,* the science of the salon.

(*This map has been coloured in the style of the period.*)

Joseph or James Moxon, 1691

THIS MANY-FACETED MAP WAS one of a series designed by the London publishers Joseph and James Moxon to be bound up with Bibles, although they were sold as separate publications too. The world map has been plotted on a north polar projection to match the star chart of the northern heavens, thus forming a twin-hemisphere design of a very unusual kind. The borders are illustrated with Biblical scenes, the upper group representing the seven days of the Creation, the lower group showing seven vital events from Biblical history, from the Fall of Adam to the New Jerusalem.

Throughout the sixteenth and seventeenth centuries, world maps had been overwhelmingly secular in purpose and in design, with religious motifs all but unknown. It seems strange therefore that an age of science should produce a world image with such a strong religious element. In fact the scientific revolution of the years 1650–1750 did not prove hostile or damaging to Christianity. Some of the greatest scientists, such as Newton and Boyle, were deeply committed to the Christian faith, and published religious as well as scientific works. The principal reason for this apparent paradox was the renewed force which the new science appeared to give to the teleological argument for God's existence — the argument from design. The empirical revolution of the seventeenth century had produced a strongly mechanistic science: observation and experiment were thought to reveal the laws by which nature and the universe functioned. The discovery and analysis of these all-pervading physical forces created a strong belief that the universe was a designed mechanism, and hence that it had a designer. As a ship is clearly designed for sailing, a blade for cutting, the eye for seeing or a bird's wings for flight, all these things, natural or human, imply the existence of a designer. Thus it was argued that it was inconceivable that the vast and complex mechanism of the solar system should exist without a designer. Boyle published in 1690 a work entitled *The Christian Virtuoso,* in which he argued that the scientific study of nature was a religious duty. Using the popular contemporary image of a clock, Boyle believed the universe to have been set in motion by the Creator, and to function now according to divine laws, whose study was the subject-matter of science. Thus the results of rational scientific study were considered to strengthen religion. In Pope's words: 'God said Let Newton be, and all was Light'.

The result of this rational-religious ethos was the curious revival of a type of *mappa mundi,* where the world is displayed together with images of its creation, and of the salvation-history of mankind. The potential hostility of science to religion did not arise, at least in England, until much later, and maps on this pattern continued to be published throughout the eighteenth and nineteenth centuries.

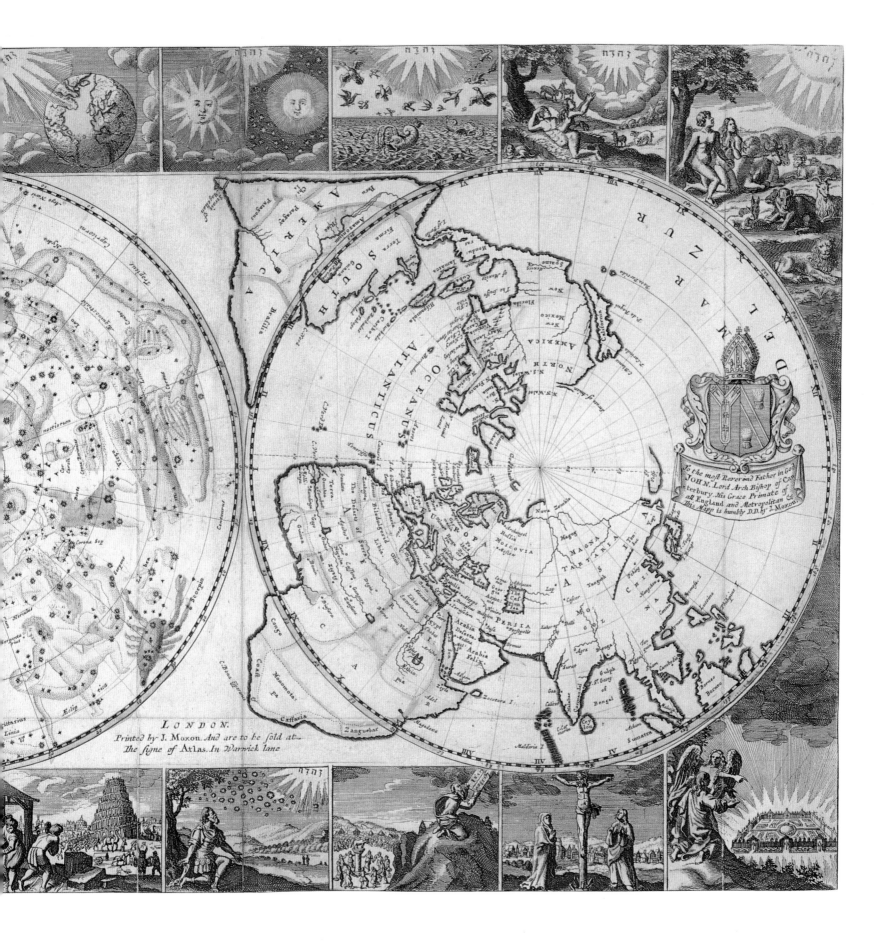

Adam Friedrich Zürner's World Map, c.1710

THIS ELOQUENT AND COMPLEX MAP, with its star charts and no less than twenty-six subsidiary hemispheres, is a visual encyclopedia of eighteenth-century geography and astronomy. Although not the earliest map of this kind — the taste for the scientific in maps appears to have begun in the 1670s — this is one of the most complete and satisfying. Adam Friedrich Zürner (1679–1742) was an important and prolific Saxon cartographer, much of whose work was published by others, often without Zürner's name. Consequently his name is less well known than his contemporaries Homann, Seutter and Schenk, and it was Pieter Schenk of Amsterdam who published this map in his atlases from c.1710 onwards. In Zürner's novel approach to the double-hemisphere world map the gods and goddesses, sea-monsters, emperors and allegorical figures are replaced by scientific, geographical information, presented in both graphic and textual form. Text and image are strikingly combined in the elaborate vignettes depicting seven natural phenomena: volcanoes, earthquakes, tides, the Norwegian Maelstrom, the winds, rain and rainbows. In the accompanying text there appears the latest date on the map (which bears no publication date), describing an eruption of Pico de Teide on Tenerife in December 1704. The world map itself shows an enlightened restraint in its avoidance of geographical myth and guesswork. Antarctica is totally blank, and the relationship between *Nova Hollandia, Diemens Land* and *Nova Zeelandia* is not artificially smoothed over. The northern Pacific however baffled all the mapmakers of this period, as islands or mainland coasts were briefly visited, then later rediscovered or renamed. *Terra Esonis* must have been a duplication of the Japanese name Yezo, the modern north island of Hokkaido; while *Compagnie Land* claimed for the Dutch East India Company, has never been identified, and might for example have been any of the Kuriles islands. The twin hemispheres are embellished with a wealth of geographical data: climatic zones, ocean currents, monsoons, and the great voyages of discovery are all charted in considerable detail. There are additional descriptive texts, such as the note on the mythical Atlantic island of *Friesland* — 'either a fable or now submerged', or that Australia was 'said by Dampier to be of all the regions of the world the most miserable'.

The celestial charts are based on the work of the great Polish astronomer Hevelius, and some of the constellations have a special contemporary interest. It became accepted practice in the seventeenth and eighteenth centuries for astronomers to designate new constellations, often with nationalistic or Christian subjects, to add to or replace those inherited from classical science. In Zürner's northern star chart for example, is an image of the shield and cross of Jan Sobieski, Poland's king from 1674 to 1696, deliverer of Vienna from the Turkish siege of 1683. Zürner's patron, Frederick Augustus, was his successor on the throne of Poland, and this constellation, first designated by Hevelius, was no doubt a symbol of national pride in Poland and Saxony. Three subsidiary hemispheres depict theories of planetary motion, the rival modern theories of Copernicus and Tycho Brahe, and the classical system of Ptolemy.

Zürner's map truly deserves the epithet encyclopedic, for there is no aspect of contemporary geography and astronomy that is not dealt with. In devising this multi-hemisphere design, the cartographer exhibits one after another the various aspects of the earth and the heavens, demonstrating the completeness of modern knowledge. In this concentration on science, Zürner's map illustrates the move away from the artistic concept of the seventeenth-century world map, to a newly-conceived world image fitting for the Age of Enlightenment.

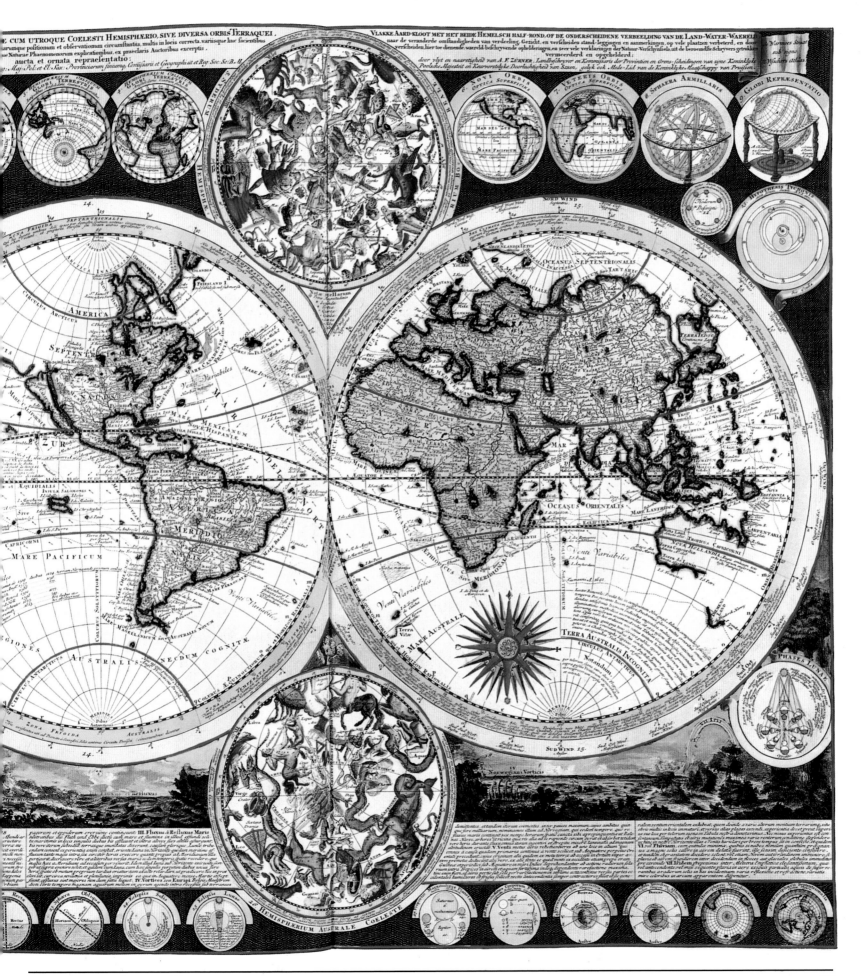

ADAM FRIEDRICH ZÜRNER, c.1710

107

Gerard van Keulen, c.1720

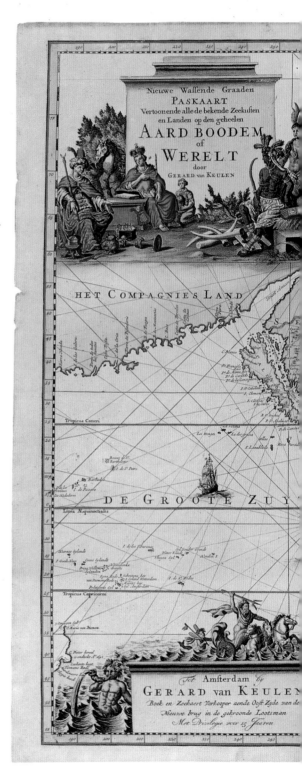

'I COULD WISH ALL SEAMEN would give over sailing by the false plain-charts, and sail by *Mercator's* chart, which is according to the truth of navigation; but it is an hard matter to convince any of the old Navigators from their method of sailing by the plain-chart; show most of them the Globe, yet they will talk in their wonted road.'

These comments by the English navigator Sir John Narborough around the year 1680, demonstrate how long Mercator's world map had to wait before its value was fully appreciated. The mathematical principles on which it was drawn, and its consequent property of showing lines of bearing as straight lines (see above, Mercator's map of 1569) were simply not understood by the great majority of working seamen. In the seventeenth century, the sailor's skill and knowledge were exclusively practical, and theoretical geography played no part in his training. More than a century after Mercator's revolutionary world map of 1569 was published, sea-charts were still being produced on the plane-chart model, ignoring the problems of global projection and directional plotting. The underlying reason for this was no doubt that before c.1600 most seamen worked exclusively in European waters, and navigated with coastal charts. A century later, navigating across oceans meant being out of sight of land for weeks or even months, and presented problems on a quite different intellectual level. In this context the virtues of the Mercator chart achieved belated recognition.

From the late sixteenth century, Dutch mapmakers had issued the most authoritative atlases of sea-charts, and the publishing house of van Keulen continued this tradition with their series *Zee-Fakkel* ('Sea-Torch', various editions from 1681 onwards). Gerard van Keulen (fl.1704–1720) was hydrographer to the Dutch East India Company, a position of great importance, with responsibility for collating all navigational data reported by Dutch seafarers, and editing new charts. In all the maritime countries of Europe, van Keulen's charts enjoyed an unrivalled position of authority during the eighteenth century, and their consistent use of the Mercator projection established it as the standard model for all sea-charts. Its authority was also recognised by publishers of general maps, and it became obligatory to include a Mercator world map in all good atlases, alongside the more familiar twin-hemisphere world map.

The long passage of time before the Mercator world map established itself as the professional, maritime world map, was caused by a dichotomy between objective science and inherited skill, between technical innovation and craft conservatism. The Mercator map's acceptance in the early eighteenth century is a further sign of the new spirit of science. A new generation of seafarers, many in government service, some with an academic education, for whom voyages across the Atlantic or the Pacific were almost routine, demanded increasingly scientific maps, maps which were truly instruments of navigation, not merely conceptual images of the world. The technical practicality of the map became the great source of its authority in a scientific age. Its geographical distortions, most notably the scale increase towards the poles, were of no practical importance for navigation. It was an ironic accident that it came to be universally accepted as an authoritative geographical image. From the time of this van Keulen map onwards, a process began in which the Mercator map came to be seen as the world map, precisely because it was the map of the seaman, the navigator, the professional world traveller.

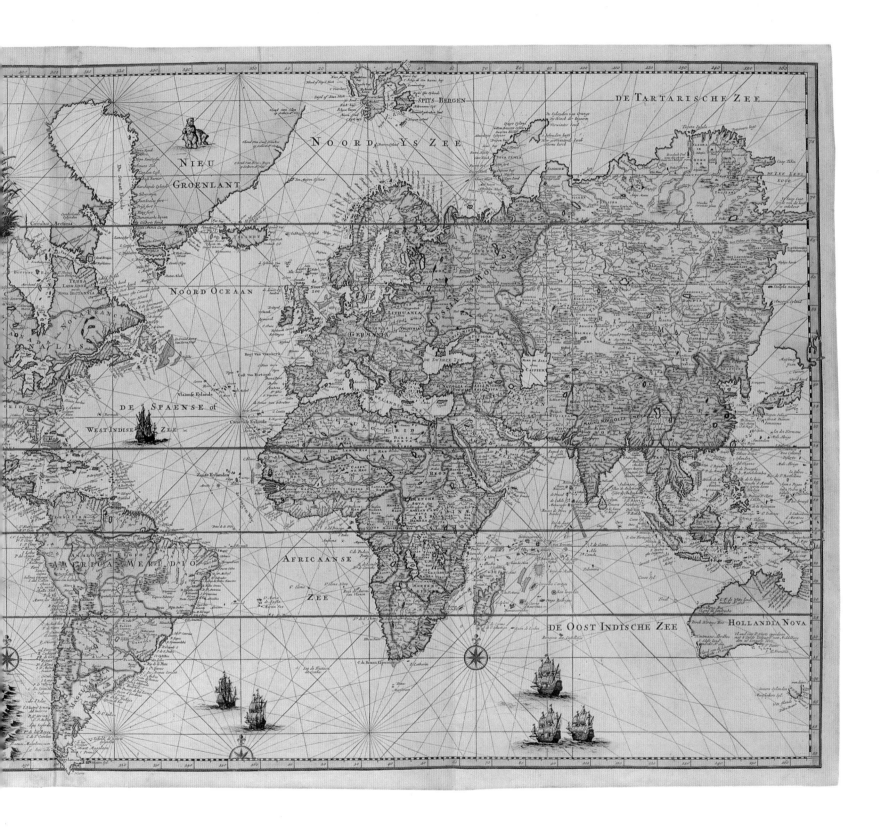

GERARD VAN KEULEN, c.1720

John Senex, 1725

THIS MAP REPRESENTS THE COMPLETE ascendancy of scientific taste in the eighteenth-century twin-hemisphere world map: the map's borders are filled neither with classical motifs nor even with scientific motifs, but with scientific texts, long and detailed passages from two of the foremost scientists of the day. Newton is represented by his theory of tides (extracted from *Principia Mathematica*), which the laws of gravitation were able to rationalize for the first time. A connection between the moon and the tides had long been conjectured, but Isaac Newton's analysis of the mass, motion and distance of the moon and the sun, accounted for a host of tidal phenomena and variations that were previously inexplicable. The other major texts transcribe from the *Philosophical Transactions* of the Royal Society Edmond Halley's accounts of the Trade Winds and Monsoons, and of the Hydrological Cycle. These phenomena are also plotted in some detail on the map, alongside a worldwide plan of magnetic compass variations. Halley, famous as an astronomer, was in fact one of the pioneers of physical geography based on observation. A two-year voyage on board a British warship in 1698–99 furnished him with the observational data that enabled him to plot lines linking points of equal magnetic variation. The problem of compass variation was a long-standing one for international shipping, and Halley's research was an early example of the conscious application of the new scientific methods to solve matters of practical importance. Halley first published his findings in a chart of the Atlantic Ocean in 1701, in a pioneering use of thematic mapping, where the map becomes a vehicle for the display of analysed data. The text here in the Atlantic reads:

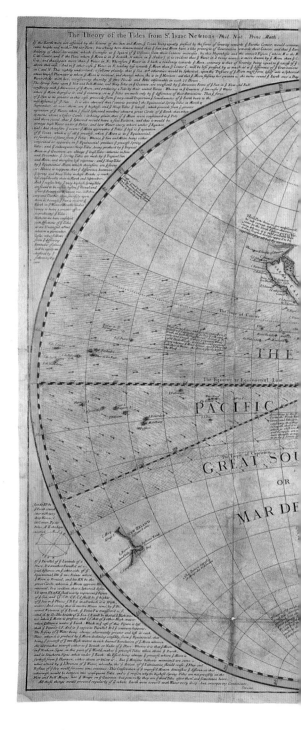

'These curved lines which express the variations of the magnetic needle were observed by Dr. Edmond Halley in the year 1700, but it must be noted that there is a perpetual though slow variation almost everywhere, viz. about C. Bona Esperanza, the westerly variation increases about a degree in 9 years, in our channel a degree in 7 years.'

For all its scientific character, Senex still includes four small figures symbolizing the continents, and most surprisingly in the South Atlantic there is a text retelling the story from Herodotus that a Phoenician fleet circumnavigated Africa around the year 500 B.C. Whatever innovations appeared in world maps, there was apparently a strong counterbalancing impulse which sought to affirm continuity with the past through the use of traditional symbols, images or concepts, appearing and re-appearing over centuries. The appearance of many explanatory texts on this map, and of the long extracts from Newton and Halley, reveal the still restricted diffusion of new scientific ideas. The map cannot stand alone without a lengthy commentary because the cartographic language is too new and contemporary interpretative skills too weak. The design of a legend or key on a map, using conventional signs to render the map self-sufficient, was just emerging at this time, and such keys obviously presupposed familiarity with the concepts or features included. Most maps from the Age of Science still found it necessary to combine text with image, just as the medieval *mappae mundi* had. Senex's map is wholly typical of the early eighteenth-century taste for scientific information in visual form; it inhabits the same world as the contemporary collections of prints of architecture, engineering, and natural history. Significantly, the map's dedicatee is Lord Burlington, the architect and immensely influential arbiter of neo-classical taste. The merely decorative aspect of the world map is now in eclipse: the aim is to inform the mind rather than to delight the eye. The map establishes its credentials as a serious scientific document, not as aesthetic decoration.

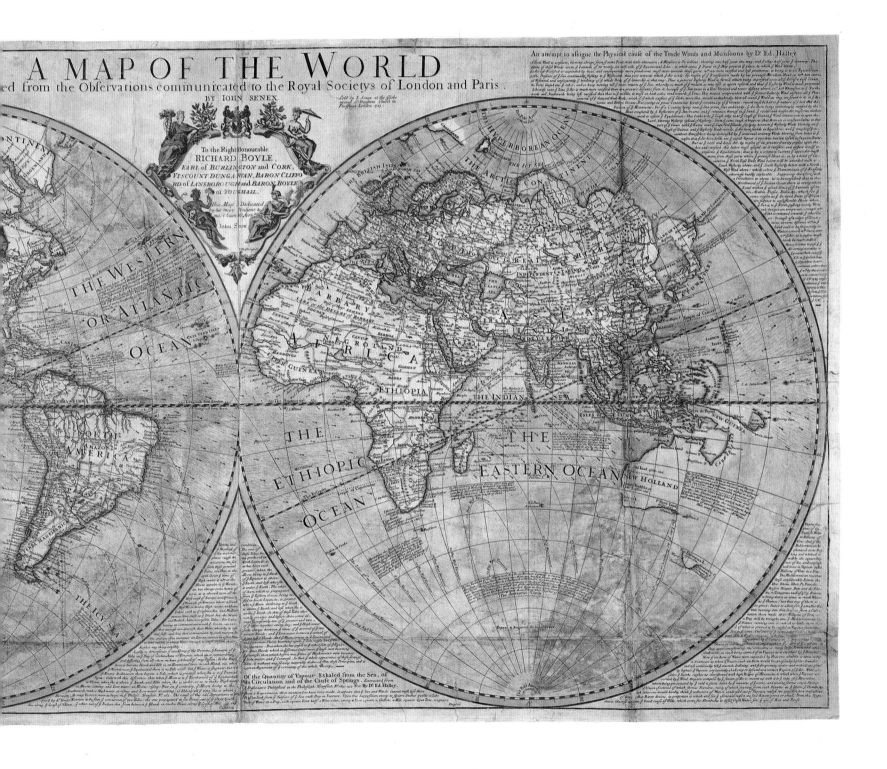

A MAP OF THE WORLD

ed from the Obfervations communicated to the Royal Societys of London and Paris .

BY JOHN SENEX

To the Right Honourable
RICHARD BOYLE,
EARL of BURLINGTON and CORK,
VISCOUNT DUNGARVAN, BARON CLIFFO
RD of LANSBOROUGH and BARON BOYLE
of YOUGHALL.

An attempt to afsigne the Physical caufe of the Trade Winds and Monfoons by Dr Ed, Halley

THE HYPERBOREAN OCEAN

THE ARCTIC CONTINENT

THE BRITISH ISLES

THE WESTERN OR ATLANTIC OCEAN

BARBARY

THE DESERT OF BARBARI

A F R I C A

NEGRO LAND

GUINEA

ETHIOPIA

ASIA

INDIA

THE INDIAN SEA

NEW GUINEA

THE EASTERN OCEAN

NEW HOLLAND

SOUTH AMERICA

THE ETHIOPIC OCEAN

THE ICY SEA

Of the Quantity of Vapour Exhaled from the Sea, of its Circulation and of the Caufe of Springs.

Joseph Delisle, 1760

IN EIGHTEENTH-CENTURY SCIENCE ASTRONOMY held centre stage. To identify and measure the force or mechanism powerful enough to move the earth and all the stars became the supreme quest to be pursued by observation and intellect. The very nature and extent of space was redefined by the telescope and by Newtonian dynamics. Much of the detailed work of eighteenth-century astronomers was concerned with the testing and confirmation of Newton's theories. In particular the three bodies sun-earth-moon presented the most obvious challenge to those who wished to calculate precisely the mass, distance and velocity within a gravitational system. Not until the works of La Place (published 1798–1827) was the entire solar system finally shown to be a stable self-supporting dynamic system. During the years intervening between Newton and La Place, there were certain important opportunities for experiment and verification, such as the predicted return of Halley's comet in 1759, and the event anticipated in this map, the transit of Venus across the Sun on 6 June 1761.

The map's publisher Joseph Delisle wrote in a pamphlet accompanying the map that 'This event has been long awaited by all astronomers for some forty years, since Halley predicted that from it we might calculate the precise distance between the sun and the earth, provided that it could be observed from the East Indies near the Ganges Mouth, and in Hudson Bay'. Halley had predicted in 1716 that Venus would appear in transit across the sun for sixteen or seventeen minutes longer at Hudson Bay than in India. The method to be used here was a parallax calculation, and Halley went on to demonstrate that the dimensions of the entire known solar system would also be obtainable. Delisle was clearly correct in emphasizing the importance of this event, since the contemporary scientific press reported at length that observers were travelling to many parts of the world months in advance to prepare for it. This map appeared in April 1760, fully fourteen months before the event, and several other map publishers in England and France exploited the great interest it aroused.

The red ellipse on the world map shows the areas of the earth from which the entire transit would be visible, with times (given in Paris time). The yellow areas would see part of the planet's course and its exit. The green areas would see the entry but not the whole course, and the white areas would see nothing at all. Despite a cloudy and frustrating morning for watchers in London, the transit was eagerly observed by astronomers all over the world, who hastened to publish their accounts of it. Some eighteen months would elapse before the worldwide data could be correlated and the necessary calculations made, then the results began to be published, giving the most accurate values yet obtained for the sun's parallax, and the consequent distances of its satellites.

The interest of this map is twofold. It shows the map publisher responding to contemporary events and shaping the intellectual taste of his time. This is an aspect of the democratization of the world map, a theme which would become still more evident in the next century. On another level it presents the image of the earth as a piece of scientific apparatus, a component in a system of celestial mechanics. The three-dimensional counterparts to this map were the tellurians and orreries which were made in the eighteenth century, and placed in libraries along with the globes and atlases. This detached, mechanical image of the world, with no hint of decoration, no echoes of classical literature and no images of secular power, was a scientific interlude between the rhetorical illustrated maps of the seventeenth century and those of the nineteenth.

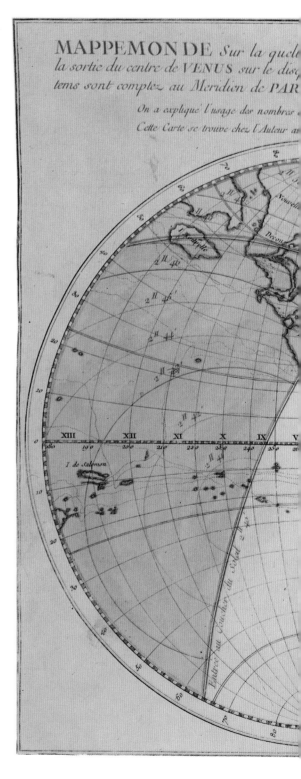

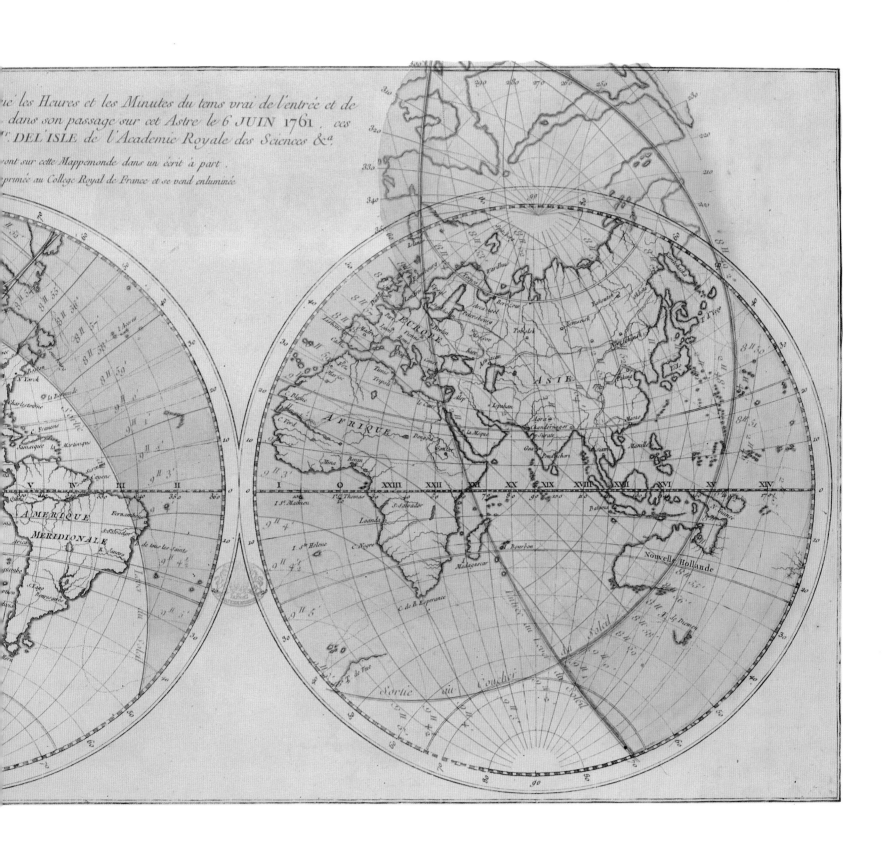

Aaron Arrowsmith, 1794

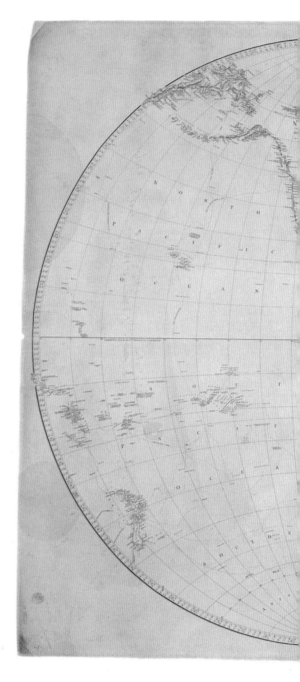

IN THE MID-EIGHTEENTH CENTURY, more than two hundred years after Magellan's great voyage, the vast area of the Pacific Ocean, approximately a quarter of the earth's surface, was still uncharted. In an age of science, this was regarded as intolerable. The possibility of a great southern continent in the Pacific still haunted the European mind, and it was hoped that the band of islands lying between 10 degrees and 20 degrees South in the Pacific would prove to be the fringes of a vast land-mass. Seeking new spheres of trade and colonization, France, the dominant European power, had formed new trading companies for India, the Pacific and China between 1680 and 1710, and a succession of exploratory voyages followed. The map of the Pacific was not readily clarified however, due to the vast distances involved, the dubious locations of earlier landfalls, the frequent 'rediscovery' of islands, and the anxiety to claim new territories. The British, watchful of French ambitions, decided in the 1760s to furnish an official Pacific expedition in a conscious effort to solve its mysteries.

'There is reason to imagine that a continent or land of great extent may be found to the southward...of the tract of any former navigators...' With these words James Cook received the official instructions from the Admiralty for his first voyage in 1768. Over the next eleven years Cook made three great voyages in which he explored more of the earth's surface than any other single man in history. He circumnavigated and charted the coasts of New Zealand and of Eastern Australia; found the strait separating Australia from New Guinea; discovered Hawaii and charted the myriads of island groups in the South-West Pacific; sailed to 71 degrees South, almost within sight of Antarctica, in the search for the southern continent; charted the west coast of Canada and Alaska, then passed through the Bering Strait to 70 degrees North, seeking a polar route out of the Pacific. His voyages were planned and executed with unprecedented care and skill, and were fully documented by Cook himself and by the scientists and artists who accompanied him. Although he narrowly failed to touch Antarctica, he proved conclusively that the Pacific held no inhabitable undiscovered continent.

This map by Aaron Arrowsmith is the fullest and clearest of the many publications inspired by Cook's achievements. Arrowsmith had access to detailed official sources through Alexander Dalrymple, the first hydrographer of the British Admiralty, whose grim portrait appears at the foot of the map. The dedication of the map to Dalrymple is deeply ironic, since Dalrymple tenaciously clung to his belief in a southern continent, a myth totally exploded by Cook. Dalrymple had hoped to lead Cook's first expedition, and he subsequently attacked Cook in print. The tracks of Cook's three journeys are clearly marked, distinguishing outward and homeward routes, with their dates. There are many precise notes drawn from Cook's records, such as that at 71 degrees South, 'Firm fields and vast mountains of ice', or that by one of the Fijian Islands 'Here Viscount de Langley and ten sailors were massacred in December 1786'. Despite the Mercator world chart's established position as the navigator's map, Arrowsmith has used the twin-hemisphere projection here, since he needed to show high latitudes in both the Arctic and Antarctic, a practical impossibility with the Mercator projection. Despite Cook's achievements, many obscurities in the Pacific remained: Japan was still veiled in secrecy, and among the minor mysteries Arrowsmith marks an island 500 miles due north of Hawaii named *Maria Lajara*, with the note that it was discovered in 1781 and was 'well inhabited'.

Comparison of this map with that of van Keulen seventy years earlier reveals the progressive demytholigizing of the world map. Together with Neptune's chariot, vast imaginary lands have finally vanished from the oceans. For the nineteenth-century explorers, there remained the polar regions and the continental interiors. For the nineteenth-century mapmakers, innova-

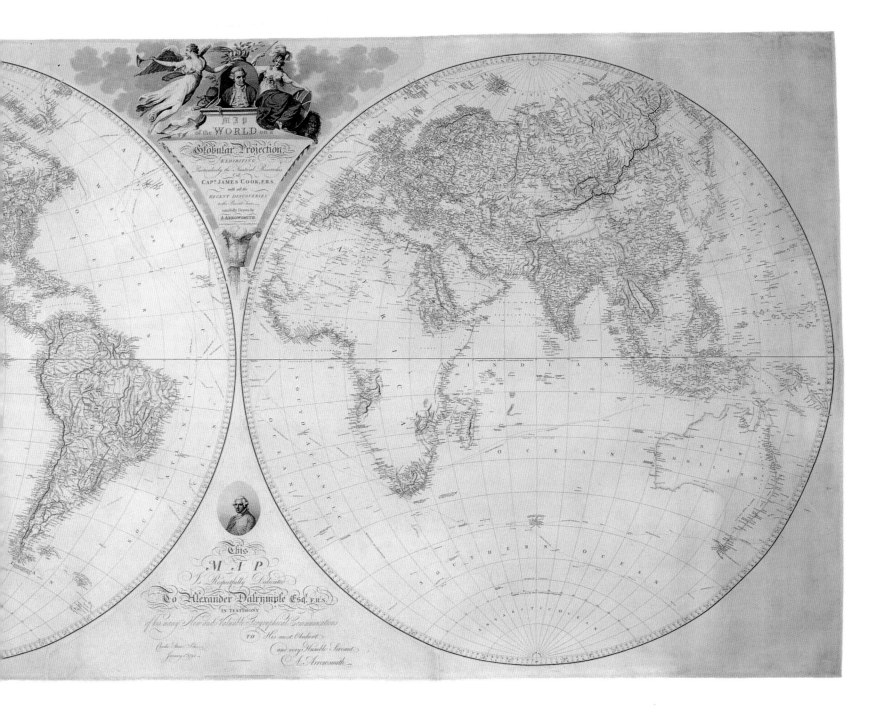

tion in world mapping would lie in its use as a vehicle for wider scientific or cultural themes. There are no final endings to geographical knowledge, but Cook's achievement specifically marked the completion of a seafaring enterprize which began almost three centuries earlier, when Dias sailed around the Cape into the Indian Ocean.

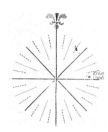

Clark,

1822

THE EARLY YEARS OF EUROPEAN overseas power were remarkable for the inhumanity with which native peoples were treated and their cultures dismissed. In time however a more ethical approach evolved, and the Evangelical and Pietist movements in late eighteenth-century Europe raised the question of the responsibilities of Christians overseas for the moral and spiritual welfare of people under their jurisdiction. The missionary movement from the 1790s onwards attracted the serious and the devout to Asia and Africa, and contributed an important element to the developing rationale of empire and responsible government. In 1813 a change in the East India Company's charter permitted Christian missionaries to work in India for the first time, and by the 1820s there were hundreds of mission stations, schools and hospitals supported by European missionary societies. A demand quickly arose for maps charting their worldwide activity, and this is one of many similar publications which appeared in this period. Nothing is known of its author, and even his Christian name does not appear on the map.

The geography of the map is peculiarly sketchy even in its picture of Europe, but accuracy was not its prime purpose. This map in effect charted the spread of European concepts and values, and the opportunity was taken to include a number of cultural indices, in order to show the context in which the missionaries were working. The full title of Clark's map claims to exhibit *The Prevailing Religion, Form of Government, State of Civilisation and Population of each Country*. The categories into which the world's religions and states of civilisation were classified were inevitably subjective and culturally conditioned; they could not be otherwise, and this constitutes the map's value as a historical document. The whole of China, India and Japan are described as pagan and half-civilized. The world's religions recognized here are Christian, divided between Roman Catholic and Protestant and Greek Orthodox; Mohammedanism; and Pagan. The faiths of Asia are not individually distinguished or dignified as serious religions. European understanding of Hinduism and Buddhism in 1822 was in its infancy. Vedic scriptures had been translated into English and German and had already created some impact as literature. The linguistic importance of Sanskrit was well-established (a professorial chair in Sanskrit was founded in Oxford in 1832), but it cannot be said that Hinduism was understood or seriously regarded by Europeans. Buddhism was at this time dimly perceived as a sect in regions such as Burma and Nepal, but its teachings and history were unknown.

The significance of this map is that it shows cultural subjectivity claiming to be an objective, semi-scientific basis for thematic mapping. The five degrees of civilization Clark identifies — *Savage, Barbarous, Half-Civilsed, Civilised, Enlightened* — correspond not to any objective criteria, but to their approach to the European model. Such maps — and they were published in many editions throughout the nineteenth century — inevitably created an image of the world that was filtered through the lens of evangelical Christianity. The age which saw the birth of scientific geography was equally the age of communication: the world map's role as a vehicle of non-geographical ideas and images was to develop as richly as at any period in the past.

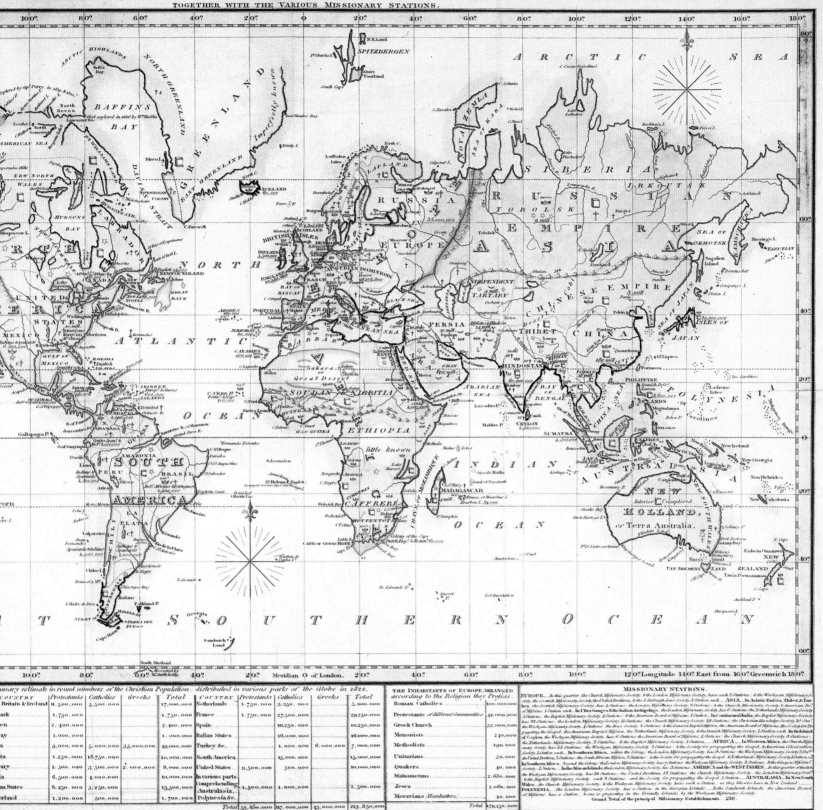

CLARK'S CHART OF THE WORLD,

...AILING *RELIGION*, THE FORM OF *GOVERNMENT*, STATE OF *CIVILIZATION*, & THE *POPULATION* OF EACH *COUNTRY*.

TOGETHER WITH THE VARIOUS MISSIONARY STATIONS.

James Reynolds: World Geology, 1849

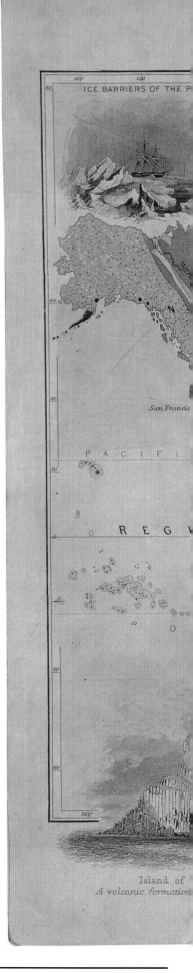

I<small>F THE QUINTESSENTIAL SCIENCE OF</small> the eighteenth century was astronomy, that of the nineteenth century was surely geology. The perception that the origin of the world and the progress of life upon it might be read in the rocks and the earth, and the claim that this new science reached back through an unsuspected immensity of time, combined to send shock waves through nineteenth-century thought and belief. They compelled a revaluation of the harmony between science and religion that had prevailed during the Enlightenment. A practical awareness of the different characters of rocks and minerals had of course existed since ancient times, but their systematic classification became possible only with the growth of analytical chemistry in the later eighteenth century. The probability that the earth is formed in strata laid down in a time-sequence had been suggested by one or two individuals as early as the seventeenth century, together with the theory of the organic origin of fossils. These three elements — mineralogy, stratigraphy, and palaeontology — permitted the publication in the early nineteenth century of the first geological maps, where colours were used to indicate the distribution of rocks and soils. The first such map of an entire country — William Smith's map of England and Wales — appeared in 1815. The great Victorian geologist, Sir Charles Lyell, brought the new science to maturity with his works *Principles of Geology* (1830–33), in which he adduced massive evidence to support the principle of inferring past events from rocks. The rapid development of geology from this point onwards, and the flow of information from field geologists in America and Asia as well as Europe, made the project of constructing a world geology map seem attractive and, almost, possible.

The London publisher James Reynolds had several unusual maps to his credit: he published a geological atlas of Britain and a number of astronomical charts. His map is colour-keyed into five major geological eras, with examples of rock types: Volcanic (greenstone, porphyry), Primary (quartz, granite), Secondary (chalk, sandstone), Tertiary (clay), and Alluvium (sand, gravel). Some of these terms have now been replaced by respectively Pre-Cambrian, Palaeozoic, and Mesozoic. Tertiary remains, while Alluvium is a term outside of geochronology. The distinction between volcanic rocks e.g. porphyry, and primary e.g. granite, has not stood the test of time, both being comprehended within the term igneous rocks. The advance and retreat of the sea was known by this time, as were some of the aspects of climate change: a text on the map informs us that 'Coal of the oldest formation was found at Melville's Island, and the plants of the coal formation of Baffin's Bay are similar to those which now flourish in the tropics'. The map also illustrates and comments on coral reefs, ice barriers in polar seas, and volcanic islands.

A considerable degree of generalization, licence, or perhaps sheer imagination must have been used in compiling this map, for there was certainly no geological data or analysis available for huge areas of Africa and central Asia, which were still unvisited by Europeans in the mid-nineteenth century. An authentic world geology map would not become possible for a further half century. Reynolds' map is nevertheless an intriguing attempt to re-map the earth in the light of contemporary science. It carried thematic mapping into a new realm of possibilities, revealing aspects of the world quite invisible to the human eye.

OLOGICAL MAP OF THE WORLD.

The north Polar Regions consist chiefly of primitive and transition rocks, with few secondary and alluvial and slight tertiary strata. Coal of the oldest formation was found at Melville Island, and the plants of the coal formations of Baffins Bay are similar to those which now flourish between the tropics.

Coral reefs are the work of organic beings which exist in inappreciable numbers. They consist of agglutinated skeletons of departed races of polypi, composed of carbonate of lime, cemented into hard calcareous rock.

CORAL REEFS

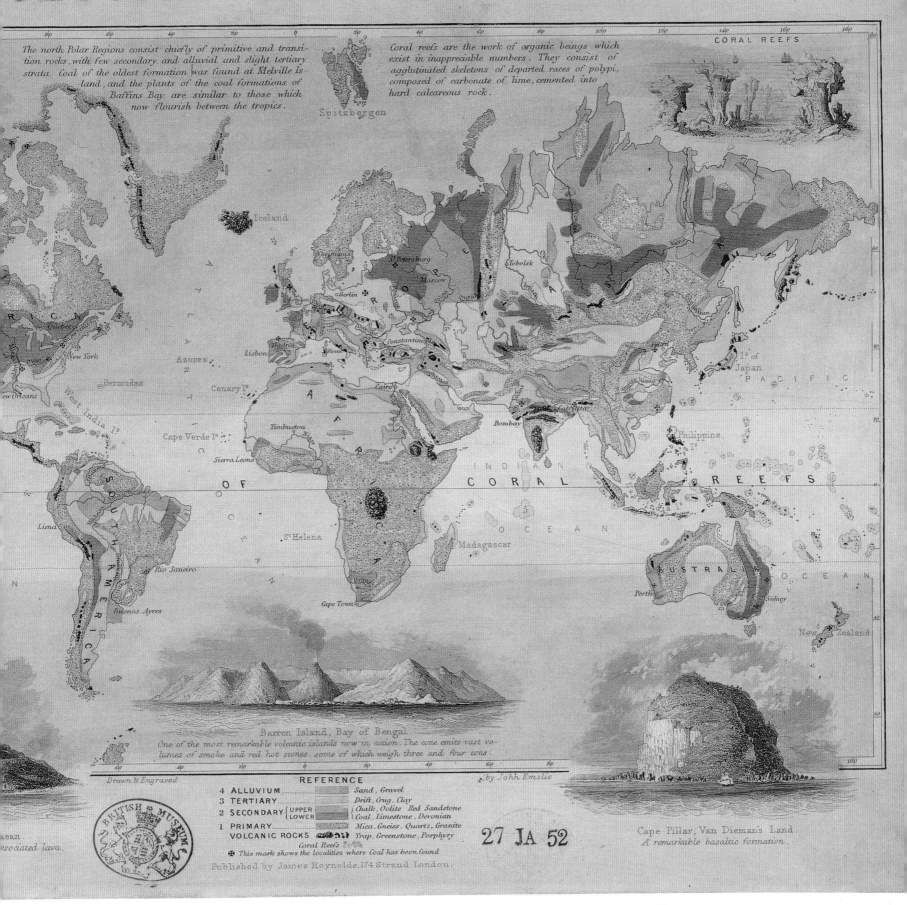

Spitzbergen

Iceland

Barren Island, Bay of Bengal.
One of the most remarkable volcanic islands now in action. The cone emits vast volumes of smoke and red hot stones, some of which weigh three and four tons.

Drawn & Engraved

REFERENCE

by John Emslie

4 ALLUVIUM		Sand, Gravel
3 TERTIARY		Drift, Crag, Clay
2 SECONDARY	UPPER	Chalk, Oolite, Red Sandstone
	LOWER	Coal, Limestone, Devonian
1 PRIMARY		Mica, Gneiss, Quartz, Granite
VOLCANIC ROCKS		Trap, Greenstone, Porphyry

Coral Reefs
✳ This mark shows the localities where Coal has been found

27 JA 52

Cape Pillar, Van Dieman's Land.
A remarkable basaltic formation.

ssociated lava.

Published by James Reynolds, 174 Strand London.

Heinrich Berghaus, 1852

AS FAR AS GEOGRAPHIC COMPLETENESS is concerned, little development remained for the world map after the mid-nineteenth century. The polar regions were the only areas of the map where significant discoveries were still to be made. But as if to balance this completion of one phase of the world map's evolution, a new category of world map appeared which offered infinite possibilities for future development. Thematic maps were designed to illustrate not the physical features of the world, but the distribution of specific phenomena — meteorology, human population, languages, volcanoes, commercial activity, and countless other subjects. The availability of large quantities of research data, scientific and human, made possible the plotting of such maps.

The most thorough and wide-ranging achievement of early thematic mapping was the series of maps by the German publisher Justus Perthes which appeared between 1845–52 under the rather imprecise title *Physical Atlas* — today we would probably call it an environmental atlas. The author of this seminal work, Heinrich Berghaus, was a Professor of Applied Mathematics in Berlin, an associate of the great von Humboldt. These maps carried out his conception of depicting cartographically all important aspects of man and his environment. His atlas consisted of some eighty maps, within the major fields of climate, hydrology, geology, natural history, and anthropology. Each map was a digest of a considerable body of information, which was elaborated in detailed accompanying texts. A number of these maps were the first such attempts to express these themes in cartographic form. Of course the highly scientific appearance of these maps was often no more than that — an appearance. Pre-suppositions of various kinds and deficiencies in data dictated the categories and the plotting of the maps. This is particularly evident in this map of the ethnic divisions of the human race. The central map, showing the principal geographic races, is only the centrepiece for a series of detailed maps and graphs showing population density, type of diet, seasons of births and deaths, life expectancy and so on.

Berghaus makes no overt moral distinctions between the races as such, but that this is a European-centred view of man's comparative strengths and weaknesses is made explicit in one of his later maps, where 'mankind's spiritual development' is charted through a system of shading: white areas in Europe indicate the highest state of civilized human life, shading through varying degrees of grey to total darkness for the unenlightened heathen regions of the world. The implications of his link between the study of ethnology and that of moral development are all too clear. It was partly as a result of anthropological data such as this map that a vital debate would arise whether the major geographic races of mankind had a common origin or had developed independently. Berghaus does not yet anticipate that question, but he adds a revealing comment to this series of anthropological maps (with an allusion to the political upheavals of 1848), which makes clear the moral purpose behind his geographical analysis:

' From these maps it can easily be seen that the human spirit must endure many more centuries before Christian truths and values, in their humanity and maturity, become the common property of mankind. That would indeed be a rich field for cartographic representation: to map the material, spiritual and moral culture of European and American peoples and lands. But reasons of time and space have decreed a halt to this work, at the moment when the European world, in its unceasing pursuit of intellectual and political progress, finds itself in the midst of an upheaval in state and society, from which future generations will assuredly benefit.'

Berghaus's new generation of quasi-scientific mapping was often limited in its sources and its interpretative methods. Nevertheless his *Atlas* was frequently re-published, and adapted into English editions, and it is historically important because it stimulated the view of the world map as a flexible model through which all manner of data, concepts and values could be communicated.

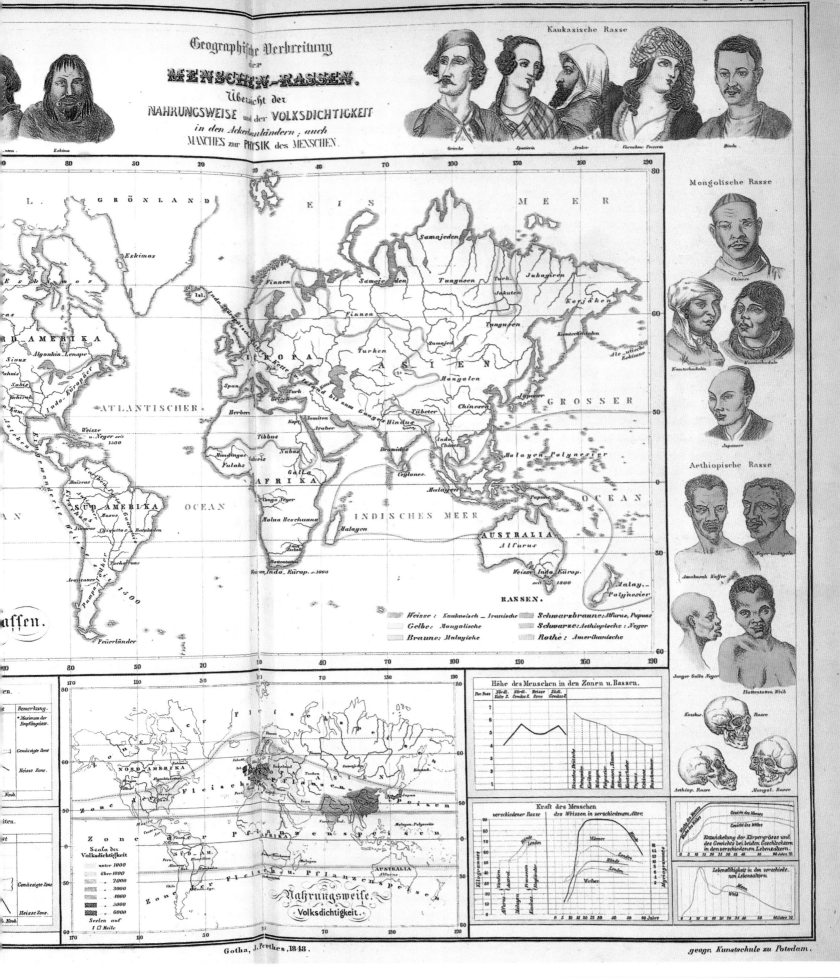

7te Abtheilung: Anthropographie N.º 1.

Geographische Verbreitung
der
MENSCHEN-RASSEN.
Übersicht der
NAHRUNGSWEISE und der VOLKSDICHTIGKEIT
in den Ackerländern; auch
MANCHES zur PHYSIK des MENSCHEN.

Kaukasische Rasse

Grieche   Spanierin   Araber   Vornehme Perserin   Hindu

Mongolische Rasse

Chinese

Kamtschadalin   Kamtschadale

Japanese

Aethiopische Rasse

Neger v. Angola

Amahosah Kaffer

Junger Galla Neger   Hottentotten Weib

Kaukas. Rasse   Malay. Rasse

Aethiop. Rasse   Mongol. Rasse

RASSEN.

Weisse: Kaukasisch – Iranische   Schwarzbraune: Alfurus, Papuas
Gelbe: Mongolische   Schwarze: Aethiopische: Neger
Braune: Malayische   Rothe: Amerikanische

Höhe des Menschen in den Zonen u. Rassen.

Kraft des Menschen
verschiedener Rasse   des Weissen in verschiedenem Alter.

Entwickelung der Körpergrösse und
des Gewichts bei beiden Geschlechtern
in den verschiedenen Lebensaltern.

Lebensfähigkeit in den verschiedenen Lebensaltern.

Nahrungsweise.
Volksdichtigkeit.

Scala der
Volksdichtigkeit.
unter 1000
über 1000
„   2000
„   3000
„   4000
„   5000
„   6000
Seelen auf
1 ☐ Meile

Zone der Fleisch u. Pflanzenspeisen

Gotha, J. Perthes, 1848.   geogr. Kunstschule zu Potsdam.

HEINRICH BERGHAUS, 1852   121

The Crystal Palace Game, 1854

THE GREAT EXHIBITION OF 1851 was an exuberant celebration of Victorian culture, reinforcing Britain's image of herself as the focal point of the contemporary world. Examples of the art, craft, science and technology from the four corners of the world were displayed to the British people between May and October, attracting six million visitors. The following year its magnificent glass and iron hall was re-erected in south London, on the site known as Crystal Palace, where it became a permanent exhibition centre. The Great Exhibition inspired many commercial products and souvenirs including several maps, and the theme of internationalism inspired this board game where players travel the world 'In search of Knowledge, whereby Geography is made easy'. The editor, Smith Evans, had also published maps of the world's shipping routes and commercial resources.

By 1854 the steamship had revolutionized intercontinental travel, and tourism, defined as travel for its own sake, was becoming fashionable. The pyramids of Egypt, tiger-shooting in India, the midnight sun in Norway, the mountains of the Cape — all were now within reach, and the wealthy began to collect destinations as well as property. Moreover, Britain's strengthening overseas ambitions, both military and commercial, meant that many thousands of people spent long periods in Africa and Asia. Few families had no contact with this wider world through soldiers, missionaries, administrators or merchants. The visual emblems of distant countries became familiar motifs in popular art — the lion, the temple, the waterfall, the pyramid. Another derivative of the Great Exhibition was James Wyld's 'Great Globe', which stood for ten years in Leicester Square: it was sixty feet high, designed to be viewed from within, its inner surface painted with the physical wonders of the world. Wyld's globe, like the Crystal Palace Game, was clearly a work of Victorian showmanship, but unquestionably they both stimulated a popular sense of geographical excitement too. This map's contemporary appeal was certainly heightened by its colour: it is an early example of Victorian chromolithography, whose vibrant colours, despite the lack of perfect register, came as a revelation in popular printing.

The Crystal Palace Game is strikingly similar to the great illustrated maps of the seventeenth century, for the visual imagery is more significant than the cartographic language of the map itself. The symbolism of the map is still concerned with power, but in a far subtler form. Here we see the power of enterprise and technology to place the world within reach of educated Europeans. Wealth and intellectual curiosity were the motivating forces of Victorian tourism, and the candid message conveyed by this map is that the world is fun and that it is available.

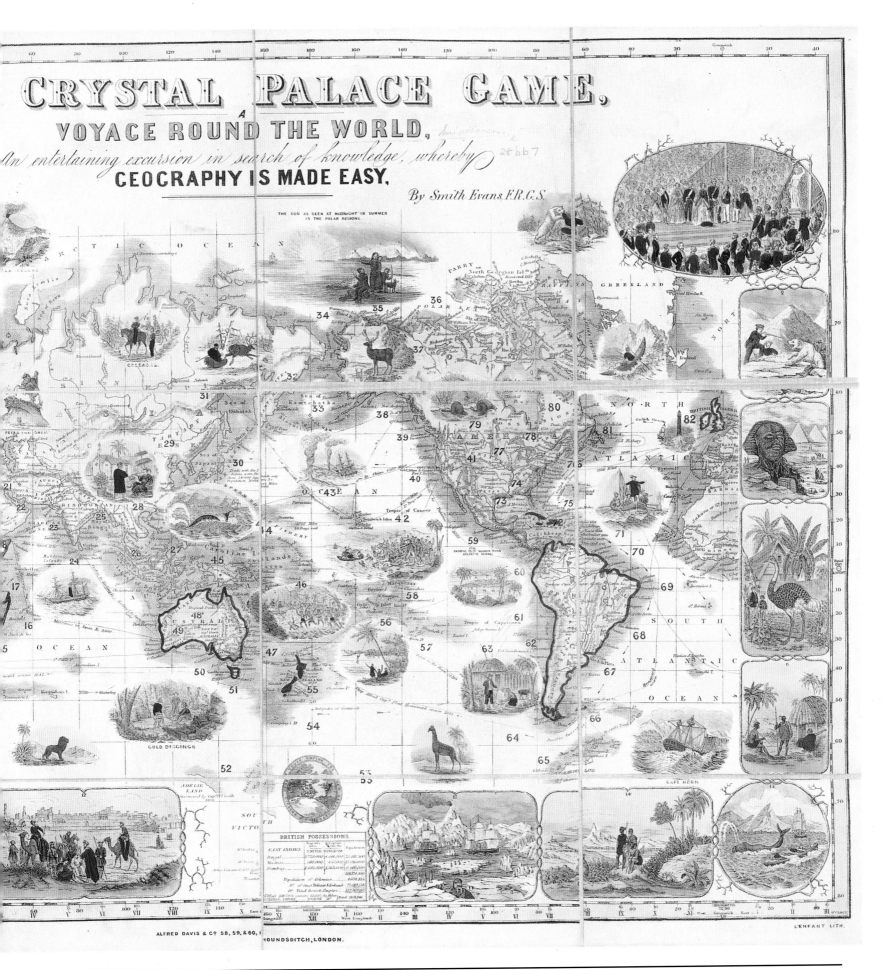

The British Empire, 1886

'TRADE HAS BEEN MADE TO flourish by war' gloated Edmund Burke after the triumph of British arms in the 1750s in India, Canada and at sea. In these overseas adventures, mainly against the French, are to be found the real origins of the British Empire. A further century was to pass however before the concept of Empire took on the mystique which made possible the creation of this map, and by that time the moving force is not war, but peace and good government. This is perhaps less a map than an icon, for rarely can a map have expressed a political philosophy so clearly. Working consciously in the tradition of the *cartes à figures* of the past, the gifted Victorian illustrator, Walter Crane, created here a cavalcade of imperial subjects, almost in the manner of a sculptural relief. Crane's art was influenced by Japanese colour prints and by art nouveau, and both these elements are clearly in evidence here. Numerous maps of the British Empire would appear in the following half-century, but none surpassed Crane's in exuberance, and if it has an element of naivety, so too perhaps did the cause it was serving.

Crane has chosen as symbolic elements in his composition the labour of the farmer and the huntsman; the fruits of earth and sea; exotic wildlife; the beauty of the noble savage; and the graces of the east. The whole tableau unfolds beneath the gaze of Britannia, her Greek helmet recalling Minerva's twin gifts of strength and beauty. While the doves of peace flutter in the map's margin, her rule is clearly guaranteed by force of arms on land and sea. The range of symbols is, to us, deeply ambivalent, but to contemporaries Britannia's blend of power and justice did not appear discordant. Crane himself was an ardent socialist, a colleague of William Morris in the Arts and Crafts movement, and even to that circle the stigma of imperialism still lay in the future.

The map was edited by Sir John Colomb, a prolific author on imperial and military subjects, and it was published by *The Graphic* magazine. The small inset map serves to contrast the extent of British territories in 1786 with the situation one hundred years later. Ironically though, had the publication been a mere decade later, the imperial red would have been more dominant still, as the regions of Africa fell like dominoes before the colonial ambitions of the European powers. Kenya, Uganda, Nyasaland, Rhodesia, and Bechuanaland all came under British control by the mid 1890's, as well as the protectorates of Egypt, Sudan, and Somaliland. The Empire, and the regions under more temporary British control, continued to expand and reached a high point by the early 1920's.

This map clearly functions on several levels — those of geography, art and politics — and so places the world map firmly in the realm of popular culture. The psychological influence of such maps can only be guessed at, but it is impossible not to believe that the elements of the whole — the map, the imagery, the concept of Empire — did not become merged in the English mind into a mesmeric image of the world and its people.

FRATERNITY · FEDERATION

MAP OF THE WORLD,
SHOWING THE EXTENT OF THE BRITISH TERRITORIES IN 1786.

WORLD

—IMPERIAL FEDERATION,—MAP OF THE WORLD SHOWING THE EXTENT OF THE BRITISH EMPIRE IN 1886.—
STATISTICAL INFORMATION FURNISHED BY CAPTAIN J.C.R. COLOMB, M.P. FORMERLY R.M.A. ——— BRITISH TERRITORIES COLOURED RED

Stanford's Library Map, 1890

THE MERCATOR PROJECTION WORLD MAP, first published in 1569, waited many years before coming into its own. The initial reason for its ascendancy in the eighteenth century was its value as a navigational chart (although its use in regional sea-charts was of more practical importance than the world map), then in the nineteenth century it came to be the most widely-used atlas and wall map. This Stanford map was not conceived as a wall map — the type is too small and the features are too finely drawn — although no doubt it was often used as one. This was a 'Library Map' in the sense that it was made to be dissected, mounted on linen, and folded into a slipcase, which then took its place on the library shelf. The publisher's catalogue gives the price as 25 shillings, a very high price in 1890 for a single map, reflecting the quality of the cartography and of this 'Library' presentation. Although the map's design is consciously severe and functional, it nevertheless achieves a cool, monumental beauty, the fluid pattern of the ocean currents counterpointing the angular land-masses.

The geography of the map is highly revealing in several areas. In 1890 the heroic period of polar exploration was just opening as the last great blanks on the modern world map awaited filling. This map extends only to 60 degrees South and to 80 degrees North, and the difficulty of charting polar exploration on this projection is clear from the ghostly land-masses which hover on the map's margins. The regional boundaries in Canada and even more strikingly in Africa, are evidence of something only too familiar to modern cartographers: a rapidly-changing political world. Comparison of successive editions of this map, and maps from the same publisher's world atlases from 1880 to 1910, will demonstrate the evolution of the Canadian regions, and the headlong scramble among European powers for colonies in Africa. The editorial processes of an international map publisher at this time were complex and secretive, and it is a pity that there is no published account of how such maps were planned, executed and cross-checked, or how international data, both political and physical, was collated. All map publishers have eagerly claimed to publish the most accurate and up-to-date maps, and commercial secrecy was no doubt important. One of the minor methodological problems in map history is this silence of the mapmakers: whether in the age of Rosselli or Blaeu or Stanford, we can only conjecture the methods and criteria by which geographical information was gathered and processed into the finished map.

Perhaps the central question that this map provokes is 'Why should the Mercator projection be so pre-eminently suitable for this type of map?' The reasons might be sought in the realms of perceptual psychology, but common sense might also provide some answers. First, because it has no artificial map edges, as all oval projections must, it fills the square sheet of paper and fits it, apparently naturally. As an extension to this, the Mercator map is a cylindrical projection which 'unrolls' the earth's surface and squares it off. Although we know intellectually that it does distort, by a sleight of hand it appears not to because it fits the square frame. Paradoxically it is the oval map which appears to distort the land-masses at its edges. Second, the unrolled cylinder seems to form a correct continuum, seen here in the double appearance of the Pacific region, whereas any oval map is strictly limited within the page. The whole area of the navigable oceans is easily encompassed, and by placing the British Isles at the centre, routes to the Far East and Australia can be studied via Suez, South Africa or Cape Horn, without interruption. Third, although it distorts sizes, it does preserve the correct shape of land-masses (technically it is orthomorphic), an important criterion on which again all oval maps must fail. Somehow it has always been intellectually more acceptable for a map to distort relative sizes than to falsify the very shape of the land.

The Mercator projection has had a bad press in recent years, but it is important to realize

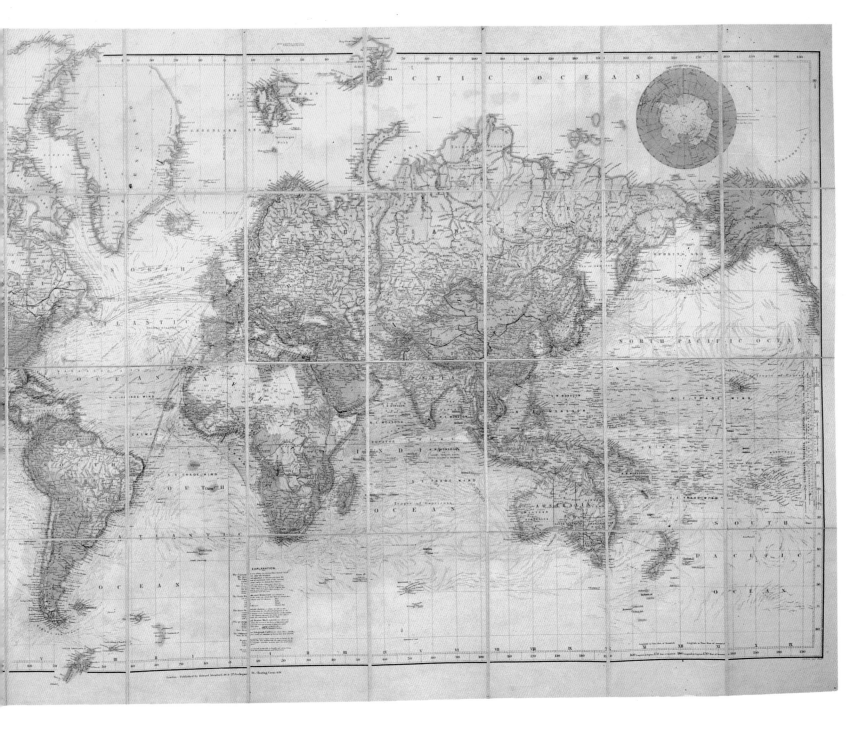

that its dominant position was neither an accident nor an intellectual conspiracy. For larger-scale regional navigational charts, it is still essential, and the construction of a world map may be seen simply as a model or paradigm of that projection. Leaving the technical aspect aside however, given that all methods of projection are compromises, the Mercator has a number of visual, practical, commonsense advantages over its rivals.

Pirate and Traveller, 1906

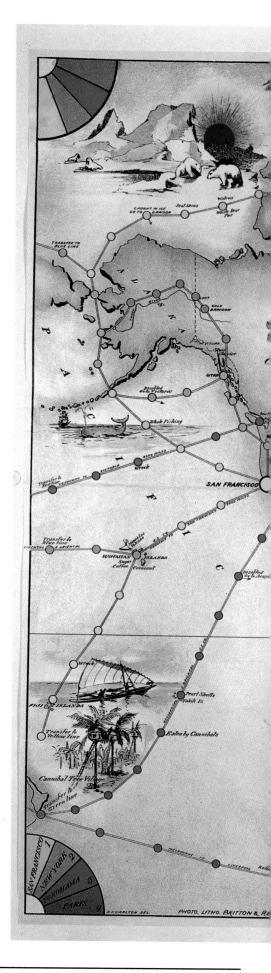

THE NINETEENTH-CENTURY FORCES WHICH made the world smaller and more accessible — steamship travel, imperialism, emigration — had an inevitable impact on the world of the child. As a specific literature for children developed, escape and adventure became recurrent themes. From *Swiss Family Robinson* and *Coral Island* through to *The Lost World, King Solomon's Mines* or *Journey to the Centre of the Earth*, a culture of exploration in an exotic but accessible world was fostered in the child's mind. This is the world of this *Pirate and Traveller* map game, in which players competed in four journeys around the world, starting from San Francisco, New York, Yokohama and Paris. Published in 1906 by the California Educational Games Company, the artwork is vivid and the geographical detail unusual. The journeys are broken by the disasters of travel — in the Atlantic 'Iceberg collision: go to Reykjavik', or in the Sahara 'Lost in sandstorm'. However 'Eaten by cannibals' has a ring of finality, and presumably signalled the end of the game for someone. The accompanying rules of the game have not survived, but presumably involved a race governed by dice throwing.

The map's educational purpose is seen in the texts describing the products of various countries, not the obvious examples of coffee from South America, but, far more originally, rugs from Teheran, opium from Bombay, camphor from Formosa and horses from Mecca. There are highly-coloured vignettes of the Aurora Borealis, of a Polynesian tree-village, and of storm-tossed ships. Conventional stereotypes of the natural world, lions and tigers, are avoided, in favour of orchids and birds of paradise. Produced for the American market, the map is Pacific-centred and shows the western United States in some detail, naming Mounts Rainier and Whitney.

The idea of using a world map in a child's game is culturally very revealing. It presupposes an educated audience with a degree of geographical awareness, and it makes explicit the link between geography and imagination. It demonstrates once again the impulse to complement the purely conceptual structure of a map with a language of visual imagery drawn from the real world. Implicit in the map is the belief that the technology and the political structures of the nineteenth century have placed the possibility of world travel within the reach of the adventurous American or European, even if they are children. It fulfills one of the enduring roles of the world map, which is to offer through a symbolic paper world, the imaginative experience of discovery. The transfer of these underlying themes into the world of the child, makes explicit the element of play that is fundamental to the creation of the world map.

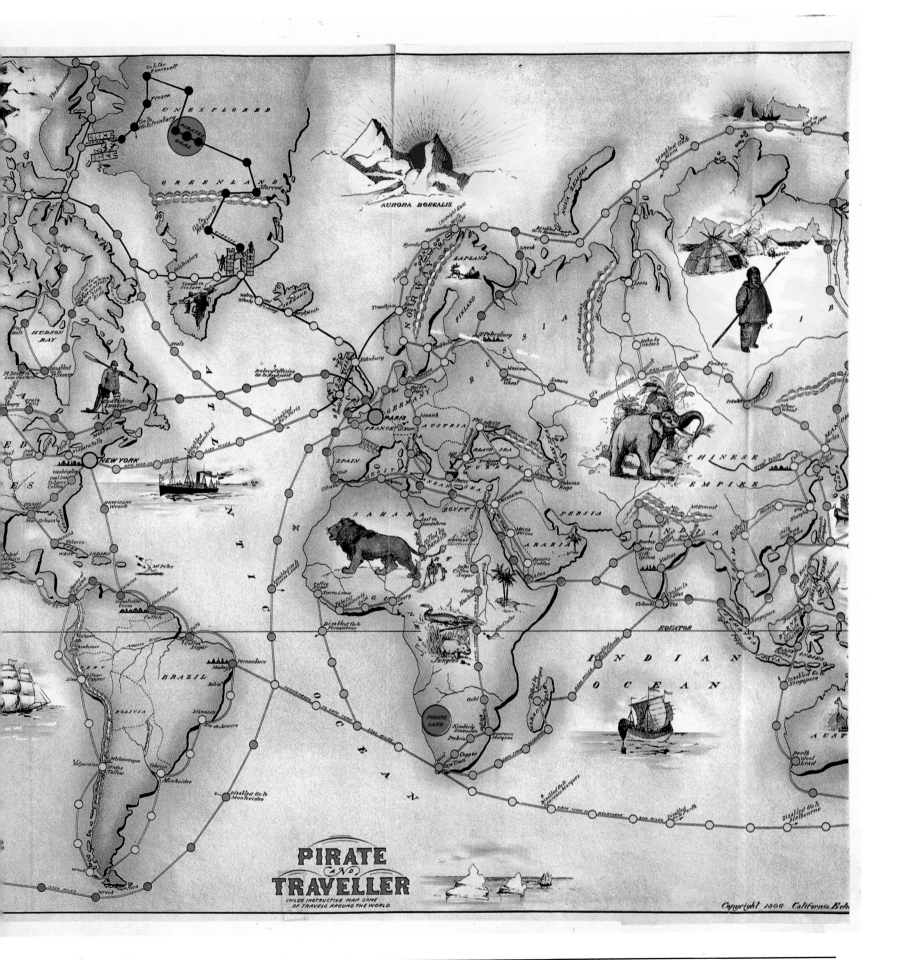

PIRATE
AND
TRAVELLER

CHILD'S INSTRUCTIVE MAP GAME
OF TRAVELS AROUND THE WORLD

Copyright 1906 California Edn

Alfred Wegener: Palaeocontinents, c.1915 (Redrawn)

MODERN GEOLOGICAL THOUGHT HAS PRODUCED, in the theory of plate tectonics, the most radically reshaped world map in the entire history of science, a map which claims, convincingly, to display the shape of the earth's land-masses millions of years before human life evolved.

The jigsaw-like fit of some continental coastlines, most obviously South America and West Africa, had been noticed in the nineteenth century by scientists such as von Humboldt. But it was the German geophysicist Alfred Wegener who pursued this to the logical conclusion that the continents had indeed once been joined and had split apart. Wegener proposed that throughout most of geological time there had been but one vast land-mass, to which he gave the name *Pangaea*, and that around 200 million years ago, in the early Mesozoic era, this had begun to fragment into the continents we know today. Common geologic formations across the now separated continental coasts added weight to Wegener's hypothesis, but because the 'new' world map was so alien, and because no conceivable mechanism of continental drift could be imagined, it received little credence for many decades. By the 1960s however, new discoveries, especially the phenomenon of sea-floor spreading, were brought together in the theory of plate tectonics, which was recognized as precisely the missing mechanism of continental movement, although the motive force remains unknown.

Plate tectonics has proved the most significant conceptual revolution in earth science. Since Wegener's original work was published in 1915, detailed research has resulted in the drawing of a complex sequence of palaeocontinental maps, from the early Cambrian era, c.550 million years ago, to the present. The migration of the earth's magnetic poles, traceable from ferrous crystals in rock, has played an important part in reconstructing this map series. These images of multiple worlds which vanished millions of years ago, have an uncanny and disturbing power, undermining the apparently fixed and permanent structures of the external world. They are maps which no geographer or scientist of the past could have conceived, although perversely they reach back to the medieval *mappae mundi*, in which one island-continent was surrounded by an encompassing sea. Map-making is here crossing the threshold of time and of human experience, mapping not that which can be seen or measured, but what is deduced from analysis and creative reasoning.

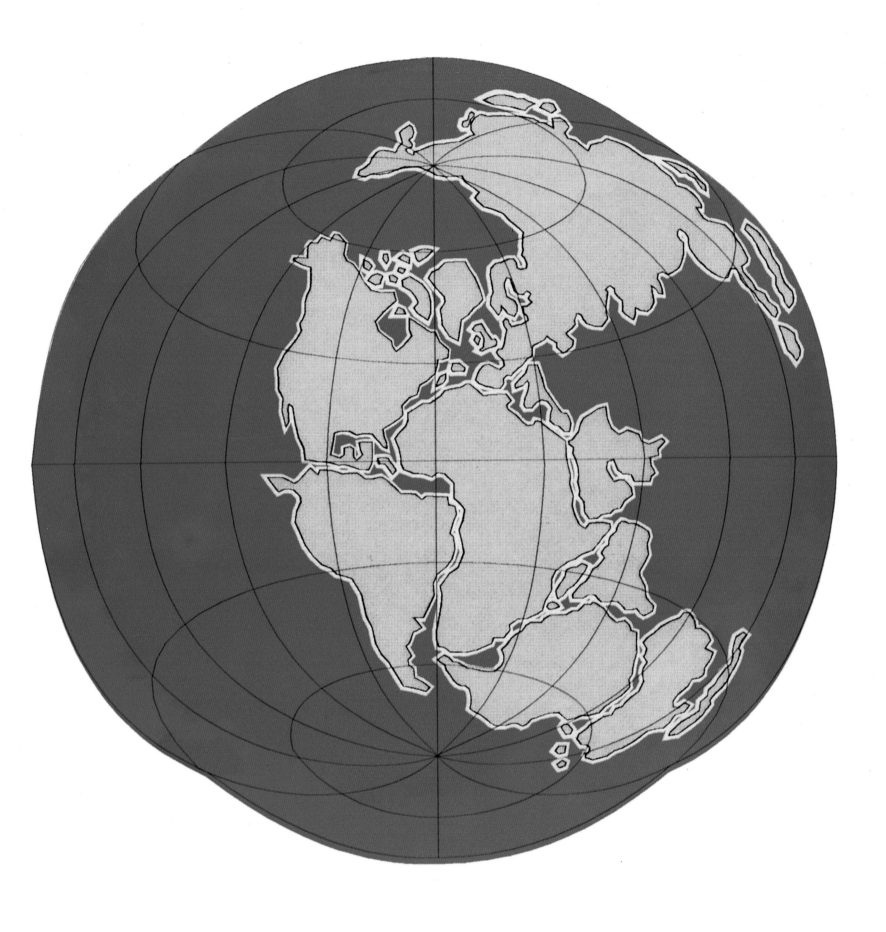

United Nations Map, 1945

I*N AUGUST 1941 WINSTON CHURCHILL* and Franklin D. Roosevelt met in conference on board warships in the North Atlantic to discuss the course of the War (in which the United States was still non-participant) and to look forward to the post-War world. The statement that was published after this meeting, the Atlantic Charter, set out the aims of worldwide security, democracy and prosperity. It was the first stage towards the birth of the United Nations, and six months later a second similar declaration was signed by twenty-six states. Building on this foundation, by October 1945, when the war had been over for just a few months, some fifty countries combined to bring the United Nations into being. This remarkable map bears witness to the optimism with which this event was seen. The whole map is charged with symbols of hope and security, the wisdom of experience and a sense of culture reborn.

The arms of the United States, Britain and the Soviet Union are merely the most prominent among those of the fifty founding members. In addition to printing the aims of the Atlantic Charter itself, extended quotations are given from the Old and New Testaments, from Lincoln's Gettysburg address, from Tolstoy's *The Kingdom of God is Within You* — a strange choice since its Christian anarchism is hostile to all political programmes. The borders of the map carry two series of vignettes of human culture in ancient and modern form: transport, communications, sport, education, industry etc. The arts of painting, drama, music and architecture are shown in sequence at the top of the map. Many countries display texts describing the key events in their history. The Axis Powers — Germany, Italy and Japan — are passed over in silence in this respect, indeed the very word Germany is omitted from the map. Numerous small banners praising famous men flutter around the map's edges, men as diverse as Moses, Socrates, Darwin, Marx and Einstein; here the giants of German culture — Bach, Beethoven and Goethe — are not omitted. In a conscious evocation of the world's diverse cultural traditions, short quotations are selected from Homer, Confucius, Shakespeare, Montaigne, and Milton. One must search to see the note of discordant contemporary reality in the tiny picture of the atomic bomb exploding over Japan.

This map is directly comparable to the great illustrated maps of the seventeenth century, with their visual language of war and peace, political power and classical culture. But this map reaches even further back in time, for it is impossible not to compare it to the medieval *mappae mundi*. It is a geographic framework which has been transformed into a cultural icon, by the display of selected contemporary emblems. Indeed it is closer to the religious *mappae mundi* than to the baroque theatre, for while secular power is strongly represented, the real message of the map is a moral one in its homage to the ideal of peace. Just as the medieval world map integrated elements of man's religious life, and stretched the boundaries of time within a geographical frame, so this United Nations map is using the image of the world to point beyond secular power to the ideal of peace, to the values of a culture which had just survived the most destructive events in world history.

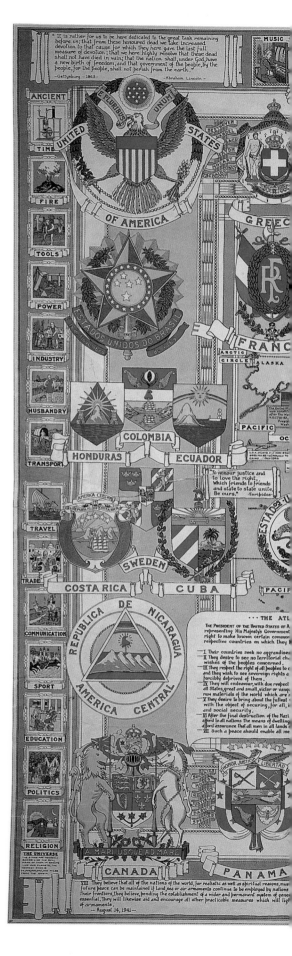

Azimuthal World Air Charts, 1946
Centering on Pearl Harbour and Delhi

T HE PROJECT OF REPRESENTING THE world on paper, of compressing three dimensions into two, has, by its very impossiblity, provoked endless explorations of the problem. There is of course no one correct and valid projection, only projections with different properties and characteristics. Since Mercator's radical approach in the sixteenth century many projections have been mathematically devised for specific purposes. The two maps illustrated here are both constructed on the azimuthal equidistant model, in which the map is conceived to touch the globe at a single point, and the rest of the world is plotted in relation to that point. The outstanding characteristic of these maps is that a straight line from the centre, from Delhi or Pearl Harbour, is the direct route and the shortest distance to any point on the globe. Moreover the scale along that line is constant. It will be seen at once that these properties would have great practical value in plotting air routes, and their principal use is in air navigation. They are the precise two-dimensional equivalent to drawing straight lines upon a globe. Their true directional properties relate only to the one central

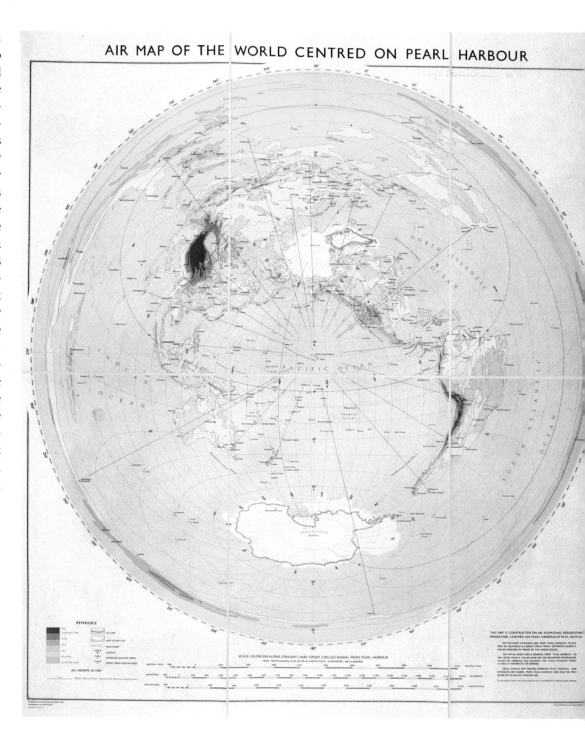

AIR MAP OF THE WORLD CENTRED ON PEARL HARBOUR

AIR MAP OF THE WORLD CENTRED ON DELHI

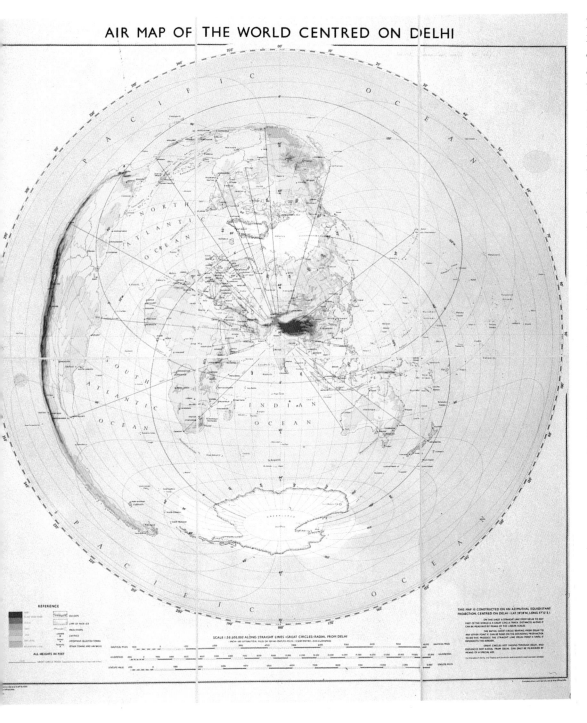

point, and a different map must be plotted for London, New York, Melbourne or any other city. The entire earth is shown, but distortion away from the central point becomes extreme, so that the resulting map is alien and apparently nonsensical. In fact it is no nearer to or further from geographic reality than for example the Mercator projection, but the latter is visually less disturbing because it is sanctified by familiarity. A study of the graticule reveals the method of construction and thereby lessens the alienating effect of the projection. The central meridian is a straight line, unusually extended through 360 degrees — normally it is a parallel which covers 360 degrees. The other meridians are complex curves intersecting at both poles. The outer rim of the map is the antipode to the point of origin, thus a single point appears as a circle. The parallels are equally spaced along the central meridian, but develop into complex curves as they approach the poles. These maps are in fact not strictly images of the world, but graphic instruments to represent in two dimensions movement through three dimensions.

Ocean Floors, 1979

'COULD THE WATERS OF THE Atlantic be drawn off so as to expose to view this great sea-gash, it would present a scene the most rugged, grand and imposing, the empty cradle of the ocean.' Thus wrote Captain Matthew Maury, the American pioneer of oceanography, in 1855. Modern techniques of ocean surveying have confirmed his prophecy. The floors of the oceans have higher mountains and deeper canyons than any found above sea-level: the island of Hawaii rises more than 30,000 feet above the Pacific floor, while the great trenches off the Philippines plunge to similar depths below sea-level. This imaginative map of the ocean floors was compiled on the basis of millions of depth-soundings, conventional and sonar, and on analysis of seismicity and magnetism. It represents a genre of science-based art which flourished in the middle years of the twentieth century, but which has now largely been overshadowed by satellite-image mosaics.

This map has been created by the human hand and eye, interpreting raw data. Its features are therefore more distinctly drawn than a satellite image will show – the vertical scale of the continental shelf for example is highly exaggerated. It shows with great clarity the Atlantic mid-ocean ridge, first traced during surveys in the 1870s. The exploration in the 1950s of the central furrows in this ridge were crucial in establishing the fact of sea-floor spreading, which led to the general acceptance of tectonic plate theory. This Atlantic ridge rises above the sea's surface in Iceland, so that we can observe on land the dynamics of worldwide plate movement – in volcanoes, earthquakes and hot springs. The phenomenon of continental shelves was first discovered in the later nineteenth century, during reconnaissance for trans-ocean telegraph cables. The shelves were immensely significant in defining the true undersea forms of the continents, revealing the possibility of land-bridges in earlier ages, between Britain and Europe for example, or across the Bering Strait.

The artistic quality of this map is worth dwelling on, since its calculated clarity has the effect of drawing back the dark mass of the sea and revealing an unsuspected region, almost a second earth. Although it·cannot strictly be defined as a thematic map, it vividly fulfils the criteria of modern scientific mapping – to chart what is invisible to the eye, and delineate a reality that is beyond our grasp.

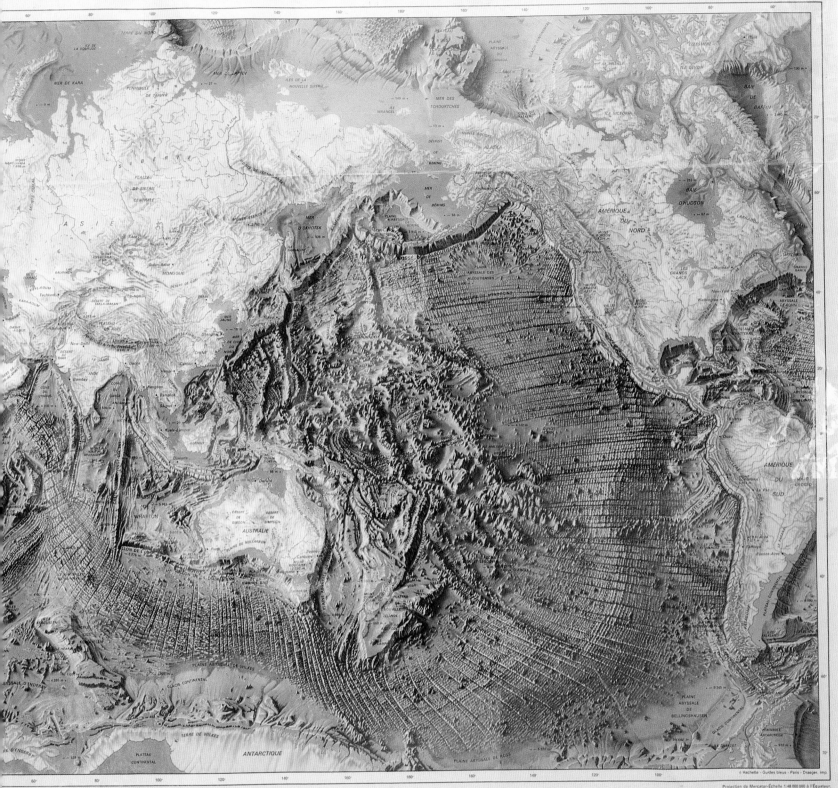

CARTE DU FOND DES OCÉANS

dressée sous la direction scientifique de
Xavier Le Pichon
du Centre National pour l'Exploitation des Océans
et d'après les relevés bathymétriques établis par
Bruce C. Heezen et Marie Tharp
du Lamont Doherty Geological Observatory
(Université de Columbia)

Dynamic Earth, 1989

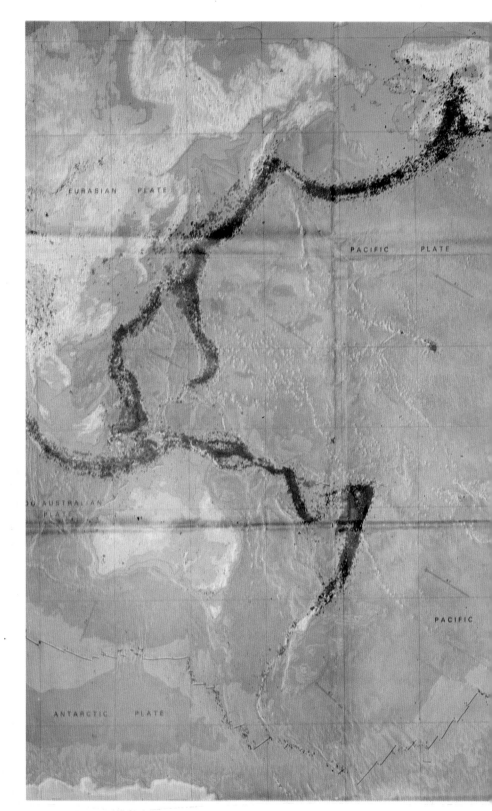

IN THE LATE TWENTIETH CENTURY, thematic mapping achieved a degree of data precision and visual impact that raised it almost to the level of an art form. The aim of thematic mapping is to locate the invisible, both in human geography and in the earth sciences. Typical of this process is the evolution of the physical map of the world. Once content to show, by the use of conventional colours, mountain, desert and forest, it can now reveal patterns ranging from the underlying structure of the earth's crust and the forces beneath it, to events in the upper atmosphere and magnetic field. The impact of this type of mapping has been to shift our perception of the earth from static to dynamic.

This forceful and persuasive map shows the earth's physiographic feature overlaid by volcanic belts, earthquake epicentres, and the outlines of the major tectonic plates. The base map is computer-generated, and shows a sensitive use of technology to create a surface texture that appears to have both depth and movement. On this base, nearly 1500 volcanoes active in the past 10,000 years are plotted, with the epicentres from 1300 significant earthquakes recorded since 1960. One of the problems unlocked by the key of plate tectonics is the location of these zones of disturbance: earthquakes and volcanoes are not random, but are concentrated along continental margins and island chains. This fact had been noticed for some time, but only the plate theory could explain it.

The general world map is accompanied by a cross-section of the earth's crust demonstrating that the apparently solid surface of the globe rides on a molten substratum. The dynamics of the plate boundaries, whether colliding or diverging, cause mountain-building, volcanoes and earthquakes. The modern scientific vision of the earth's interior bears a striking resemblance to the archetypal legends of a burning underworld, the realm of demonic forces which from time to time erupt into the world of man.

Innovative thematic maps such as this create novel forms of the image of the world. The outline of the invisible tectonic boundaries, highlighted with the menacing red of the volcanic belt, is now as familiar as the pink-tinted Mercator map of the British Empire was to an earlier generation. Thematic maps, perhaps even more than geographical maps, do not embody a single objective form of the world map, but create a diversity of images that are meaningful for their time.

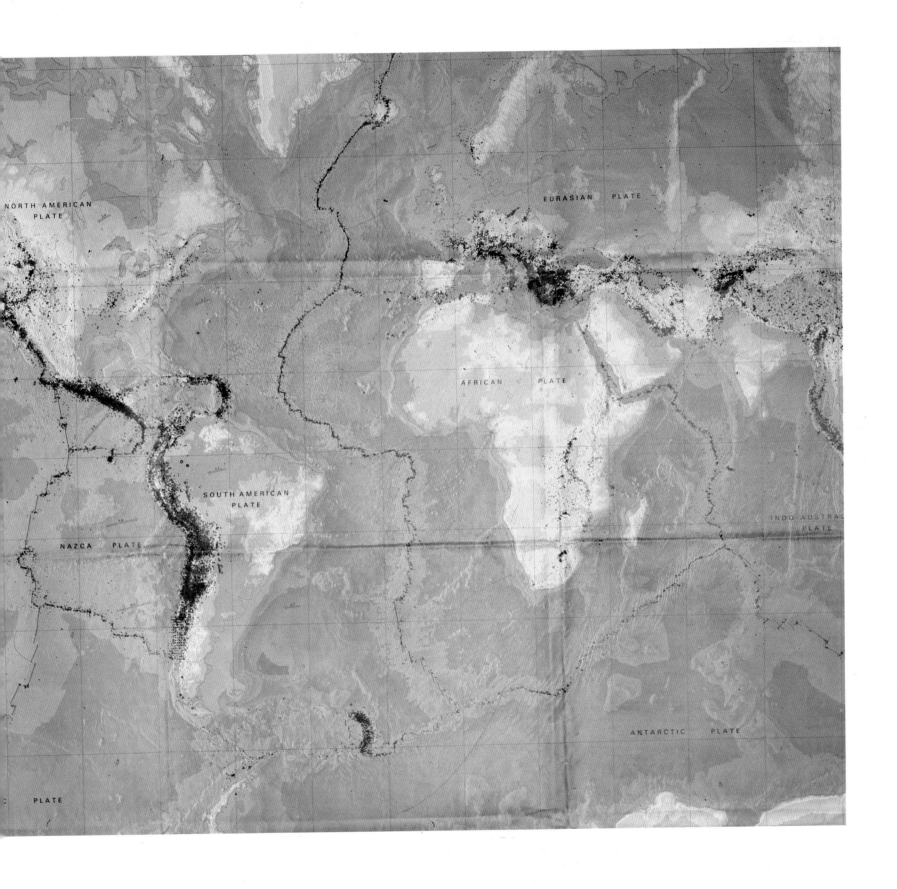

NORTH AMERICAN
PLATE

EURASIAN   PLATE

AFRICAN   PLATE

SOUTH AMERICAN
PLATE

NAZCA   PLATE

INDO-AUSTRAL
PLATE

ANTARCTIC   PLATE

PLATE

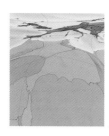

Cartogram, 2008

THEMATIC MAPPING HAS BEEN VERY GOOD at displaying large, general categories of information, for example geology, the religions of the world, or the main languages – things which could be differentiated by colours. But we now live in a world awash with information far more detailed than that, primarily social and economic data, which cannot be fitted into the mode of thematic mapping. The answer to this problem came with the development of cartograms in the 1970s, a development closely connected to the rise of computer use.

Cartograms are visual representations of quantitative data presented as a map differentiating the higher-value areas from the lower simply by size. The most obvious example would be population: a cartogram of world population would draw each country at a size relative to its population. The result would be visually very odd, but still recognisable as an image of the world. The publication of the celebrated Peters projection world map in 1973 was highly influential in the development of cartograms, for it showed the countries of the world in their correct physical areas, and offered a very different image of the world from the one made familiar by Mercator. The Peters map itself was not a cartogram, it was a projection, but it undoubtedly prepared our minds for the possibility of seeing a different world map, one redrawn to display specific data or ideas which the conventional world map did not show. It also tended to throw the spotlight on the developing world far more than conventional world maps did.

Modern cartograms are constructed by using a computer programme which relates mathematical data to a given unit or area of the map. We might think of it as pouring figures into that area, causing it to expand or contract. This effect is very evident in the style of cartogram shown here: whatever the subject that is being plotted, the area with the highest incidence grows larger, while the lowest grows smaller. In this way hundreds, potentially even thousands, of themes can be plotted on maps – although they will relate overwhelmingly to measurable social and economic indicators. One of the great advantages of cartograms is that they convey information quickly and with great impact. Here, the appallingly high infant mortality rates of sub-Saharan Africa and India leap out and hit you; in India in 2002 the infant mortality rate stood at 6.7 per 100 live births, compared to 0.7 in the United States. If a number of cartograms on themes such as this are grouped together, then geographical patterns of deprivation can be built up, literally producing a picture, which cannot be hidden or explained away. Cartograms like this are very different from traditional maps, helping us see our lives in relation to those of people in far distant countries, and see the world as a global village.

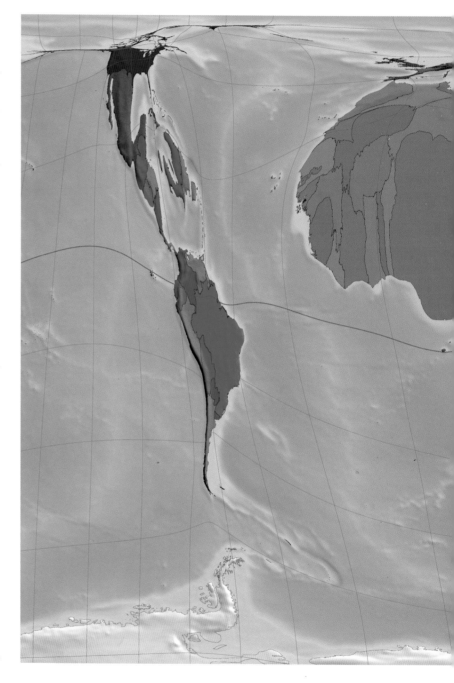

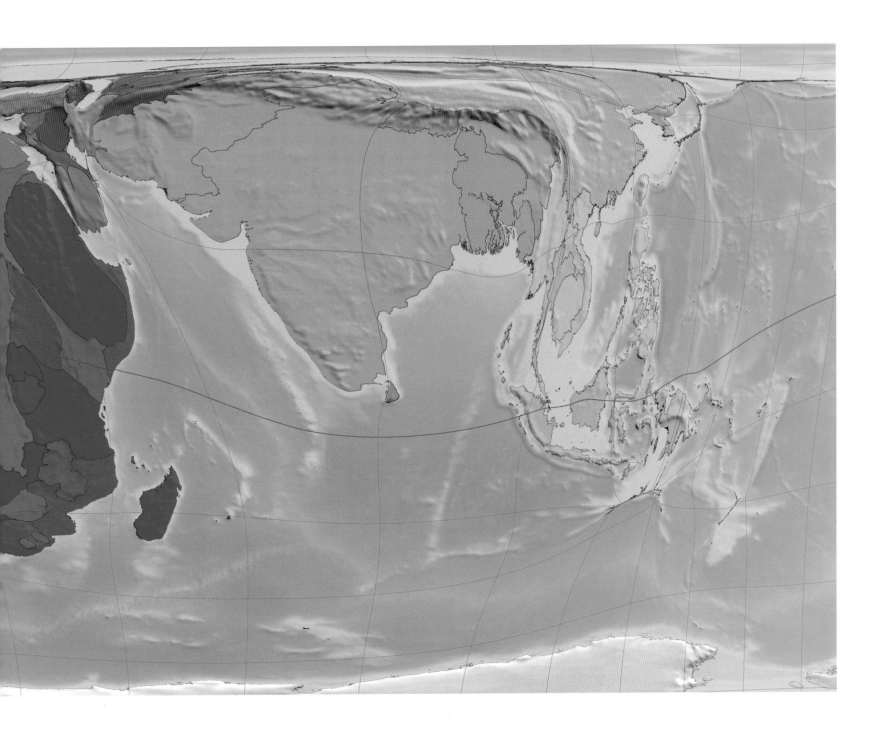

Digital and Internet Mapping

IN THE TWENTY-FIRST CENTURY, cartography, like so many other things, is in the throes of a revolution – that of digital and internet mapping. On the internet, maps and map-related materials are everywhere: some individual map websites respond to a million requests an hour. Never have maps been so much used or so familiar: after centuries when they were instruments in the hands of a few specialist geographers, explorers or military men, they have lost all the mystique or aura that they once had, and have become images on a screen, often with a life of just a few seconds. What are the implications of this revolution for our conceptual image of the world?

In the first place, maps have become far more specific and functional, touching areas of our daily lives like weather, traffic, property searches and holidays, and we expect them to be updated not just every day, but every hour, even every minute. In the second place they have become interactive: the map is no longer a frozen image to be admired or studied, instead we want it to do something, to respond to and change at our command, to reveal different levels of information. Thirdly, they are effectively free, so that we don't have to choose whether to acquire them, they are simply there, waiting to be utilised. Finally, they are ephemeral, in fact more than ephemeral, they are instantaneous and transient – they appear to come from nowhere and return to nowhere.

All this adds up to a radical democratisation of geographical data: anyone can now find a map instantly, scan it, and throw it away. This has been made possible by digitisation, the splitting of the map into tiny basic elements which can be re-assembled and manipulated at will, and onto which innumerable other kinds of data can be imposed. Maps are no longer just lines on paper, but electronic pulses, meaning that they can be manipulated or transformed, and integrated with other information. This gives immense power to the mapmaker and the map user, but it come at a price.

Firstly there is the conceptual price: fragmentation of geographical data leads to the fragmentation of our image the world. We now ask maps for one specific bite of information, we don't ask for a unified vision of the world. The internet map user can become power-hungry, excited more by the manipulation of data than by any sense of the real physical world. Nowhere is McLuhan's doctrine that 'the medium is the message' more clearly proved, and his prophecy that man first re-shapes his technologies and then they re-shape him, has come true with disturbing accuracy. To imagine that digital mapmaking is merely a technical change, invisible in the map itself, is as naïve as to argue that the invention of printing did not change the world of books.

The second price is social, and the sinister potential of the fragmented, electronic map is made clear in the military use of digital cartography. Modern warfare would be unthinkable without the weapons which are based entirely on geographical information systems, that is, on encoded maps. Away from the battlefield, internet surveillance techniques are also dependent on such maps and geographical coordinate systems, raising profound issues of freedom and civil liberties.

For centuries, traditional cartography was based on physical reality, on the surface of the earth, to which most aspects of human life were tied, and this was appropriate in societies that were static, conservative and hierarchical. The environment in which we now live is more fluid, dynamic and rootless, less certain and less permanent, and our mapping reflects this. Maps now claim to chart the way we live – our rising or declining wealth, our economic activity and resources, our social norms and problems, our zones of international conflict, our leisure habits, where we fly to, what we eat and how we die. We are consumers of information, and mapping is part of the new information economy. The intangible things – concepts, beliefs, histories, cultures, values – these do not appear in this new generation of 'real world' maps.

Obviously there can be no going back to the age of the Mappa Mundi, to a world dominated by Christ, or to the great visual maps of the seventeenth century, to the world of the merchant-adventurer, or even to the United Nations map of 1945, with its moral ideas of peace, hope and civilisation overspreading the earth. So perhaps the final gift – the poisoned gift – of the digital map age is to tell us that there is no image of the world any more, but only data. Digital mapping dissects the world in order to analyse how it works, but the maps of the past were more than data, for they showed how the imagination might re-create an image of the world that was comprehensive, living and in some fundamental way meaningful.

Further Reading

Baker, J.N.L., *A History of Geographical Discovery and Exploration*, new ed. 1963, Harrap.

Binding, P., *Imagined Corners: Exploring the World's First Atlas*, 2003, Headline Review.

Buisseret, D., *The Mapmakers' Quest: Depicting New Worlds in Renaissance Europe*, 2003, Oxford University Press.

Campbell, T., *The Earliest Printed Maps 1472–1500*, 1987, University of California Press (Berkeley).

Cosgrove, D., *Apollo's Eye: a Cartographic Genealogy of the Earth in the Western Imagination*, 2001, The John Hopkins University Press.

Dekker, E. and van der Krogt, P., *Globes from the Western World*, 1993, Zwemmer.

Dorling, D. et al., *The Atlas of the Real World: Mapping the Way We Live*, 2008, Thames & Hudson.

Edson, E., *The World Map 1300–1492: the Persistence of Tradition and Transformation*, 2007, The John Hopkins University Press.

Edson, E. and Savage-Smith, E., *Medieval Views of the Cosmos*, 2004, Bodleian Library.

Fiorani, F., *The Marvel of Maps: Art, Cartography and Politics in Renaissance Italy*, 2005, Yale University Press.

Harvey, P., *Medieval Maps*, 1991, British Library Publishing.

Heitzmann, C., *Europas Weltbild in alten Karten: Globalisierung im Zeitalter der Entdeckungen*, Herzog August Bibliothek Wolfenbüttel, 2006.

Kamal, Y., *Monumenta Cartographica Africae et Aegypti*, 5 vols in 16 parts, 1926–1951, privately published.

Mollat du Jourdin, M. et al., *Sea Charts of the Early Explorers*, 1984, Thames & Hudson.

Nebenzahl, K., *Maps from the Age of Discovery, Columbus to Mercator*, 1990, Times Books. Published in US as *The Atlas of Columbus and the Great Discoveries*, 1990.

Palmer, D., *The Complete Earth*, 2006, Quercus.

Scafi, A., *Mapping Paradise: the History of Heaven and Earth*, 2006, British Library Publishing.

Shirley, R.W., *The Mapping of the World: Early Printed World Maps 1472–1700*, reprinted 1993, New Holland Publishers.

Short, J.R., *Making Space: Revisioning the World, 1475–1600*, 2004, Syracuse University Press.

Sicilia, F., ed., *Alla Scoperta del Mondo: l'arte della Cartografica da Tolomeo a Mercatore*, 2001, Il Bulino (Modena).

Singer, C., *A Short History of Scientific Ideas to 1900*, 1959, Oxford University Press (Oxford) and Clarendon Press (New York).

Stefoff, R., *The British Library Companion to Maps and Mapmaking*, 1995, British Library Publishing.

Whitfield, P., *The Charting of the Oceans*, 1996, British Library Publishing.

Whitfield, P., *New Found Lands: Maps in the History of Discovery*, 1998, British Library Publishing.

Wolter, J.A. and Grim, R.E., *Images of the World: the Atlas through History*, 1997, McGraw-Hill.

The History of Cartography, University of Chicago Press,
Volume 1, Harley, J.B. and Woodward, D., eds., *Cartography in Prehistoric, Ancient and Medieval Europe and the Mediterranean*, 1987.
Volume 2, Book 1, Harley, J.B. and Woodward, D., eds., *Cartography in the Traditional Islamic and South Asian Societies*, 1993.
Volume 2, Book 2, Harley, J.B. and Woodward, D., eds., *Cartography in Traditional East and Southeast Asian Societies*, 1994.
Volume 2, Book 3, Woodward, D. and Lewis, G.M., eds., *Cartography in Traditional African, American, Arctic, Australian and Pacific Societies*, 1998.
Volume 3, Woodward, D., ed., *Cartography in the European Renaissance*, 2 vols, 2007.

Picture Acknowledgements

Index

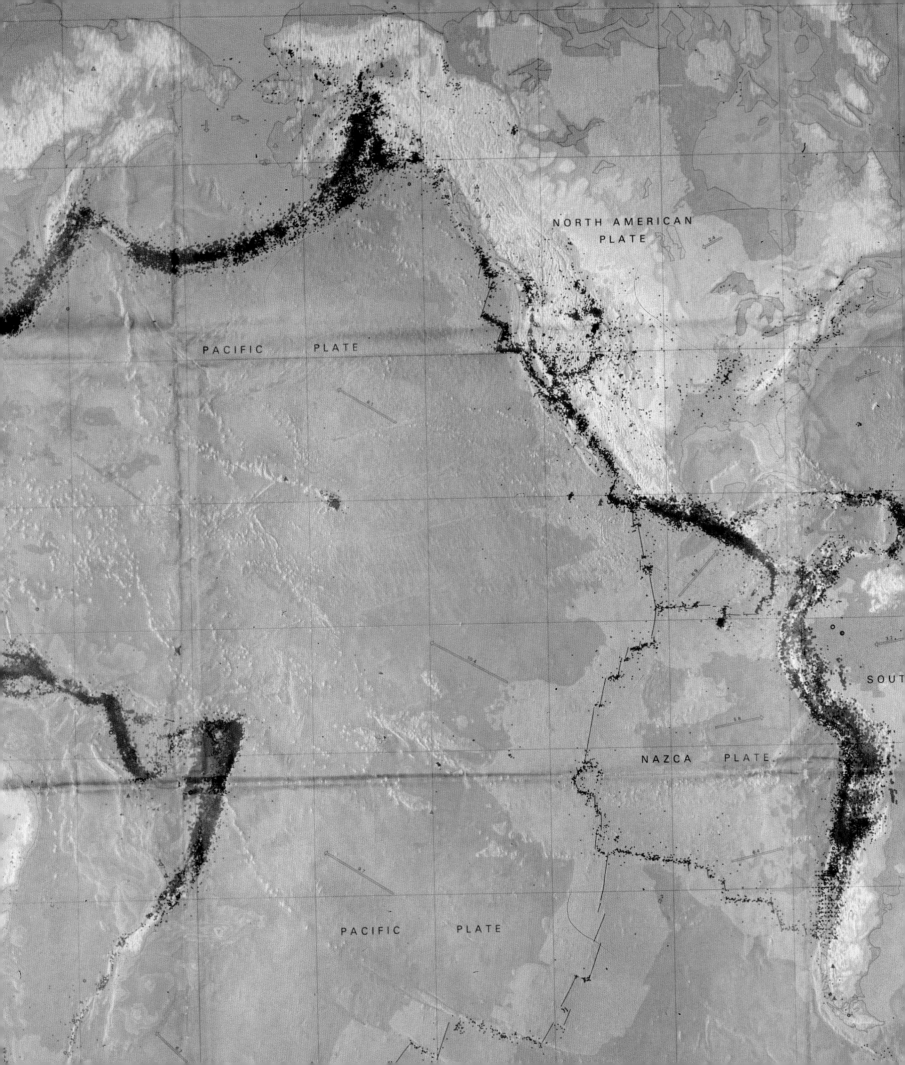